AF308368

Jörg Illig

Physically based Impedance Modelling of Lithium-Ion Cells

Schriften des Instituts für Werkstoffe der Elektrotechnik,
Karlsruher Institut für Technologie

Band 27

Eine Übersicht aller bisher in dieser Schriftenreihe
erschienenen Bände finden Sie am Ende des Buchs.

Physically based Impedance Modelling of Lithium-Ion Cells

by
Jörg Illig

Dissertation, Karlsruher Institut für Technologie
Fakultät für Elektrotechnik und Informationstechnik, 2014

Impressum

 Scientific
Publishing

Karlsruher Institut für Technologie (KIT)
KIT Scientific Publishing
Straße am Forum 2
D-76131 Karlsruhe

KIT Scientific Publishing is a registered trademark of Karlsruhe
Institute of Technology. Reprint using the book cover is not allowed.

www.ksp.kit.edu

Print on Demand 2014

ISSN 1868-1603
ISBN 978-3-7315-0246-3
DOI: 10.5445/KSP/1868-1603

Physically based Impedance Modelling of Lithium-Ion Cells

Zur Erlangung des akademischen Grades eines

DOKTOR-INGENIEURS

von der Fakultät für
Elektrotechnik und Informationstechnik
des Karlsruher Instituts für Technologie (KIT)
genehmigte

DISSERTATION

von

Dipl.-Ing. Jörg Nils Illig
geb. in Karlsruhe

Tag der mündlichen Prüfung: 04.04.2014
Hauptreferentin: Prof. Dr.-Ing. Ellen Ivers-Tiffée
Korreferent: Prof. Dr. Hubert A. Gasteiger

Danksagung

Die vorliegende Dissertation entstand während meiner Forschungstätigkeit am Instituts für Werkstoffe der Elektrotechnik (IWE) des Karlsruher Instituts für Technologie (KIT). An erster Stelle gilt mein aufrichtiger Dank Frau Professor Ivers-Tiffée für ihr entgegengebrachtes Vertrauen und das hervorragende Arbeitsumfeld. Durch ihre engagierte Förderung meiner fachlichen und persönlichen Entwicklung habe ich in dieser Zeit neben dem Erstellen dieser Arbeit viel gelernt. Weiterhin gilt mein herzlicher Dank Herrn Professor Gasteiger für sein Interesse an dieser Arbeit, seinen hohen Einsatz vor der mündlichen Prüfung und natürlich für die Übernahme des Korreferats.

An dieser Stelle möchte ich mich ebenso bei Professor Imanishi und seiner Gruppe für die Ermöglichung meines Auslandsaufenthaltes an der Mie-Universität in Japan und die entgegengebrachte Gastfreundschaft bedanken. Diese Zeit trug entscheidend zur Reife und Fertigstellung dieser Dissertation bei. In diesem Zusammenhang bedanke ich mich außerdem beim Karlsruhe House of Young Scientists (KHYS) für die finanzielle Unterstützung dieses Auslandsaufenthalts.

Meinem Gruppenleiter André Weber bin ich für die wissenschaftlichen Diskussionen und die vertrauensvolle Zusammenarbeit zu Dank verpflichtet. Ebenso möchte ich mich im Besonderen bei meinen Kollegen Moses Ender und Jan Philipp Schmidt für eine kollegiale und immer wieder motivierende Zusammenarbeit bedanken. Auch allen weiteren Mitgliedern der Batteriegruppe und der anderen Arbeitsgruppen möchte ich für die angenehme Atmosphäre und die schönen Momente abseits der fachlichen Zusammenarbeit danken. Unvergesslich werden für mich auch das freundschaftliche Klima und die Kollegialität am IWE bleiben. Stellvertretend möchte ich mich dafür bei meinem langjährigen Zimmergenossen André Leonide und meinem "Alt-Kollegen" Michael Kornely bedanken. Sie haben die Atmosphäre am IWE durch Ihre außergewöhnlichen Persönlichkeiten maßgeblich geprägt.

Meinen ehemaligen Studenten und Hiwis danke ich für die gute Zusammenarbeit und ihre stetige Unterstützung. Sie haben wesentlich zum Gelingen dieser Arbeit beigetragen. Besonders erwähnt seien hierbei meine beiden letzten Diplomanden Michael Weiss und Christian Uhlmann, die ich inzwischen auch als Kollegen schätzen gelernt habe.

Ohne technische Unterstützung würde eine experimentelle Promotion wie die meine nicht funktionieren, weshalb ich mich bei allen technischen Mitarbeitern und der Werkstatt des IWE für die zuverlässige Zusammenarbeit bedanken möchte. Auch Andrea Schäfer sei für ihre Hilfsbereitschaft in organisatorischen Angelegenheiten gedankt.

Bei meinen Projektpartnern aus dem Projekt KoLiWIn und der Kooperation mit der BMW AG möchte ich mich ebenfalls für die vertrauensvolle Zusammenarbeit bedanken. Außerdem danke ich der Firma Bosch und insbesondere Herrn Dr. Ziegler für den regelmäßigen offenen Austausch während der ersten Jahre meiner Promotion.

Meinen Freunden sei für die Unterstützung und vor allem für die so wichtige Ablenkung außerhalb des Arbeitsalltags gedankt.

Zu guter Letzt bedanke ich mich von ganzem Herzen bei meiner Freundin Carmelina für ihre moralische Unterstützung und ihr großes Verständnis während der letzten Jahre sowie bei meiner Familie für die Eröffnung aller erdenklichen Möglichkeiten.

Jörg Nils Illig

Karlsruhe, 01.08.2014

Zusammenfassung

Die effiziente Speicherung elektrischer Energie hat in den letzten Jahren durch den Ausbau der erneuerbaren Energien und der Elektromobilität signifikant an Bedeutung gewonnen. Ein aussichtsreicher Kandidat dafür ist die Lithium-Ionen Technologie, welche eine Energiespeicherung unter Bereitstellung hoher Energie- beziehungsweise Leistungsdichten erlaubt und deshalb bereits als Standardtechnologie in Laptops, Handys sowie Elektrowerkzeugen eingesetzt wird. Eine Weiterentwicklung der Lithium-Ionen Technologie ist aufgrund der gestiegenen Anforderungen an Leistungsfähigkeit, Lebensdauer und Sicherheit jedoch dringend erforderlich. Um diese Eigenschaften gezielt optimieren zu können, wird ein genaues Verständnis der physikalischen Vorgänge in Lithium-Ionen Zellen und ihren Elektroden benötigt.

Das vorrangige Ziel dieser Arbeit war die Identifikation und Quantifizierung aller relevanten Verlustprozesse im Betrieb einer Lithium-Ionen Zelle, die potentiell zu einer Beschränkung der Leistungsfähigkeit und zur Alterung führen können. Dazu wurden Lithium-Ionen Zellen und deren Elektroden mittels elektrochemischer Impedanzspektroskopie und Zeitbereichsmessungen analysiert. Die Kombination beider Messmethoden ermöglichte die Untersuchung eines breiten Frequenzbereichs von $100\,\mathrm{kHz}$ bis $1\,\mathrm{\mu Hz}$. Diese Messdaten wurden anschließend mithilfe der Verteilung der Relaxationszeiten ausgewertet, um anoden- und kathodenseitige Verlustprozesse zu separieren und deren Abhängigkeiten von Betriebsparametern (Ladezustand, Temperatur) aufzuklären.

Für die Kathode wurde aus den zahlreichen Alternativen $LiFePO_4$ als Modellsystem ausgewählt, das aufgrund seiner Beständigkeit bei erhöhter Betriebstemperatur die Sicherheitsanforderungen der Automobilindustrie am besten erfüllt. Für die Anode wurde Lithium-Metall beziehungsweise Graphit ausgewählt, das sich in kristalliner oder amorpher Struktur für alle Lithium-Ionen Zellen etabliert hat. Für die elektrochemische Charakterisierung und Modellierung von Kathode und Anode wurden sowohl kleine Experimentalzellen mit einer Elektrodenfläche von $2{,}54\,\mathrm{cm^2}$ aufgebaut, als auch kommerzielle Leistungszellen der Dimension 18650 mit $1{,}1\,\mathrm{Ah}$ Kapazität untersucht. Die Ersatzschaltbildmodelle, die zunächst für Anode und Kathode getrennt aus Experimentalzellmessungen aufgestellt wurden, wurden anschließend durch die Analyse der kommerziellen Leistungszelle auf ihre Übertragbarkeit geprüft. Alle Untersuchungsergebnisse und Modellierungsansätze werden im Folgenden im Detail vorgestellt.

Modellierung von LiFePO$_4$-Kathoden

In Kapitel 5 wurden im Labormaßstab hergestellte LiFePO$_4$-Kathoden mittels elektrochemischer Impedanzspektroskopie unter Variation der Betriebsparameter Temperatur (T) und Ladezustand (SOC, engl. *State Of Charge*) systematisch charakterisiert. Dazu wurden sogenannte Halbzellen bestehend aus LiFePO$_4$-Kathode und metallischer Lithium-Anode in einem Laborzellgehäuse gemessen. Anschließend wurden die Impedanzspektren durch die Anwendung der Methode der Verteilten Relaxationszeiten (DRT, engl. *Distribution of Relaxation Times*) ausgewertet. Diese Methode wurde am Institut für Werkstoffe der Elektrotechnik (IWE) zur Analyse von Brennstoffzellen entwickelt und in dieser Arbeit erstmalig auf die Lithium-Ionen Zelle übertragen. Zunächst war dafür die Entwicklung einer Vorgehensweise zur Vorverarbeitung der Spektren notwendig, da die Impedanzspektren einer Batterie im Gegensatz zur Brennstoffzelle für tiefe Frequenzen ein durch die DRT nicht beschreibbares rein kapazitives Verhalten zeigen. Die Vorverarbeitung beinhaltet (a) den Entwurf eines Ersatzschaltbildmodells zur Beschreibung der Festkörperdiffusion und (b) das anschließende Abziehen dieser Teilimpedanz vom Impedanzspektrum der Zelle. Dadurch und durch die Messung symmetrischer Kathoden- beziehungsweise Anoden-Zellen wurde eine Identifikation und physikalische Interpretation aller relevanten Verlustprozesse erreicht. Diese sind für die LiFePO$_4$-Kathode im Einzelnen: (i) Die Festkörperdiffusion $P_{diff,C}$ im Aktivmaterial, (ii) der Ladungstransfer P_{1C} zwischen Aktivmaterial und Elektrolyt und (iii) der Kontaktwiderstand P_{2C} zwischen Kathodenschicht und Stromableiter.
Daraus konnte ein physikalisch basiertes Ersatzschaltbildmodell vorgeschlagen werden, welches jeden dieser Verlustprozesse durch ein entsprechendes Ersatzschaltbildelement repräsentiert. Es beinhaltet für die LiFePO$_4$-Kathode zwei RQ-Elemente zur Beschreibung von P_{1C} und P_{2C} sowie ein Generalized Finite Length Warburg Element in Serie mit einer Kapazität zur Beschreibung von $P_{diff,C}$. Durch Anfitten des Modells an die Messdaten konnten die Parameterabhängigkeiten (SOC,T) jedes identifizierten Verlustprozesses bestimmt werden.
Durch die vorgestellte Vorgehensweise wurde erstmalig die Identifikation und Quantifizierung aller relevanten Verlustprozesse einer LiFePO$_4$-Kathode ermöglicht. Des Weiteren wurde erstmals die Anwendung der DRT zur Identifikation von Verlustprozessen und zur physikalisch basierten Modellentwicklung für Batterieelektroden präsentiert. Die entwickelte Vorgehensweise ist universell einsetzbar und kann somit auf jede beliebige Elektrodenchemie angewendet werden.

Optimierung von LiFePO$_4$-Kathoden

Kapitel 6 untersucht die Abhängigkeit der für die LiFePO$_4$-Kathode identifizierten Verlustprozesse von der Elektrodenmikrostruktur. Diese Abhängigkeit wurde an der im Labormaßstab hergestellten Kathode durch Variation der Parameter (i) Schichtdicke, (ii) Materialanteile und (iii) zusätzliches Verdichten (Kalandrieren) der Schichtstruktur genauer betrachtet. Dazu wurde eine Konfiguration der Experimentalzelle mit integrierter Referenzelektrode entwickelt, um eine zuverlässigere Analyse ohne den Einfluss der Gegenelektrode zu erhalten. Die so beobachtbaren Zusammenhänge bestätigten die zuvor getroffene Zuordnung der Verlustprozesse und das entwickelte Ersatzschaltbildmodell. Ein linearer Zusammenhang zwischen aktiver Oberfläche und Ladungstransferprozess P_{1C} sowie ein minimierter Kontaktwiderstand P_{2C} bei erhöhtem Leitrußanteil und zusätzlichem Verdichtungsschritt wurden festgestellt. Die Mikrostrukturparameter aktive Oberfläche, Phasenanteile und Porosität wurden dazu über eine dreidimensionale Rekonstruktion der im Labormaßstab hergestellten $LiFePO_4$-Kathode gewonnen (Dissertation Moses Ender, IWE 2014 [1]). Eine derartige Korrelation zwischen Mikrostruktur und Impedanz ist aus der Literatur bisher nicht bekannt.

Durch Variation der Entladerate unter Last konnte der Einfluss der Mikrostrukturparameter auf die Leistungsfähigkeit der $LiFePO_4$-Kathode bestimmt werden. Die im Labormaßatab hergestellte Kathode konnte so um den Faktor 2,5 in der Leistungsdichte bei Entladeraten von über 5C gesteigert werden. Trotzdem blieb die kommerzielle $LiFePO_4$-Kathode aufgrund einer für hohe Leistungsdichten "maßgeschneiderten" Mikrostruktur überlegen. Der Kontaktwiderstand zwischen Kathode und Stromableiter sowie die niedrige elektronische Schichtleitfähigkeit wurden als leistungsbegrenzende Faktoren für die labortechnisch hergestellte $LiFePO_4$-Kathode identifiziert.

Für die Optimierung von Elektroden mit höherer flächenspezifischer Kapazität und niedrigen Widerständen kann die hier vorgestellte Vorgehensweise angewendet werden. Allerdings wird dann eine Anpassung des Zellaufbaus für die Entladeexperimente erforderlich, da sowohl die Lithium-Anode als auch der Elektrolyt in diesem Fall beginnen, signifikant zur Begrenzung der Leistungsfähigkeit beizutragen. Dies kann durch Verwendung einer Referenzelektrode für die Spannungsmessung in Kombination mit einem sehr dünnen Separator erreicht werden.

Modellierung kommerziell eingesetzter Anoden

In Kapitel 7 wird die Untersuchung der aus der 18650 Zelle entnommenen Graphitanoden in Experimentalzellgehäusen vorgestellt. Die Verwendung einer Referenzelektrode erlaubte die separate Betrachtung der Graphitanode, deren wichtigste Ergebnisse im Folgenden kurz zusammengefasst werden.

Durch die Impedanzanalyse unter Variation der Temperatur und des Ladezustandes und die anschließende DRT-Auswertung wurden vier relevante Verlustprozesse für die Graphitanode identifiziert: (i) Die Festkörperdiffusion im Aktivmaterial, (ii) der Ladungstransfer und (iii) die SEI zwischen Aktivmaterial und Elektrolyt sowie (iv) der Kontaktwiderstand zwischen Anodenschicht und Stromableiter. Zusätzlich wurde der Beitrag des Lithium-Ionen Transports in den Poren der Elektrode bestimmt. Die Identifikation und Trennung dieser fünf Beitrage war mit der bisher aus der Literatur bekannten direkten Auswertung der Impedanzspektren nicht möglich.

Die Analyse der Mikrostrukturparameter und der berechneten DRTs implizierte die Verwendung sogenannter Leitermodelle zur korrekten Beschreibung der Graphitimpedanz. Diese Modellstruktur erlaubt eine Berücksichtigung des ionischen und elektronischen Pfades durch die poröse Elektrode. Der separate Beitrag jedes Verlustanteils zum Impedanzspektrum wurde durch CNLS-Fits über einen weiten Parameterbereich (T=0 °C bis 30 °C, SOC=0 % bis 100 %) bestimmt. Die ionische Leitfähigkeit im Elektrolyten stellte sich dabei für hohe Temperaturen (20 °C bis 30 °C) und hohe Ladezustände (über

10 %) als Hauptbeitrag zum Impedanzspektrum heraus. Für niedrigere Temperaturen und Ladezustände waren der Ladungstransfer und die SEI für die größten Verluste verantwortlich, da diese beiden Prozesse mit sinkender Temperatur stark anwachsen. Dies resultierte in hohen Aktivierungsenergien von 0,79 eV für den Ladungstransfer und 0,72 eV für den Lithium Transport durch die SEI. Der große Beitrag des Elektrolyten in den Poren war bisher bei kommerziellen Standardgraphitanoden in der Literatur nicht bekannt.

Anschließend wurde das für die Hochleistungsanode entwickelte Ersatzschaltbildmodell auf die Graphitanode einer Hochenergiezelle übertragen, um seine Allgemeingültigkeit für die Analyse unterschiedlicher Graphitanoden zu überprüfen. Es konnte gezeigt werden, dass das Model sowohl in seiner Struktur als auch in der Wahl seiner einzelnen Impedanzelemente ebenso zur Beschreibung der Hochenergieanode geeignet ist. Dabei wurde für letztere – wie anhand der Mikrostrukturparameter beider Graphitanoden aus der dreidimensionalen Rekonstruktion zu erwarten war (Dissertation Moses Ender, IWE 2014 [1]) – ein stärkerer Einfluss des Lithium-Ionen Transports in den Poren identifiziert.

Zusammenfassend wurden in Kapitel 7 mittels DRT-Analyse und Ersatzschaltbildmodellierung alle in einer Graphitanode auftretenden Verlustprozesse inklusive des Lithium-Ionen Transports durch die SEI und in den Poren identifiziert. Die Kombination aus DRT-Analyse und Mikrostrukturparametern sowie die daraus abgeleitete Beschreibung durch Leitermodelle erlaubt erstmalig eine gezielte Analyse des Einflusses der porösen Elektrodenstruktur auf das Impedanzspektrum von kommerziell hergestellten Anoden. Die Berücksichtigung dieser Erkenntnisse ist zur physikalisch sinnvollen Impedanzanalyse der Alterung von Graphitanoden unabdingbar.

Mehrstufige Analyse von 18650 Lithium-Ionen Zellen

Die detaillierte Analyse der 18650 Zelle mit Hilfe der zuvor gewonnen Erkenntnisse wird in Kapitel 8 vorgestellt. Zunächst erlaubte die Untersuchung der ausgebauten Elektroden in Experimentalzellgehäusen eine zuverlässige Trennung der Anoden- und Kathodenverlustprozesse mittels Referenzelektrode. Weiterhin führt der reduzierte Einfluss der Induktivität in Experimentalzellen zu einer Erweiterung des hochfrequenten Bereichs bei Impedanzmessungen von ca. 5 kHz auf bis zu 100 kHz.

Der niederfrequente Bereich wurde durch die Ergänzung der Impedanz- mit Zeitbereichsmessungen um vier Dekaden von 10 mHz auf 1 µHz erweitert. Die verwendeten Zeitbereichsmessungen wurden am IWE zur Analyse kommerzieller Lithium-Ionen Zellen entwickelt und im Rahmen dieser Arbeit das erste Mal auf Experimentalzellen angewendet.

Die DRT-Analyse führte zur Identifikation von sieben Verlustprozessen, die in der 18650 Zelle frequenzabhängig durch den Einsatz von Experimentalzellen und physikalisch motivierter Ersatzschaltbildmodelle für ein weites Feld von Betriebsbedingungen quantifiziert werden konnten:

1. Frequenzbereich I von $100\,\text{kHz}$ bis $5\,\text{kHz}$ (gemessen durch Impedanzspektroskopie): Das Impedanzspektrum der 18650 Zelle wird hier durch induktives Verhalten dominiert, welches vom Messaufbau und von der Zelle selbst herrührt. Durch Experimentalzellmessungen konnte gezeigt werden, dass nicht nur der Elektrolyt sondern auch die Kontaktwiderstände zwischen den Elektroden und den Stromableitern ($P_{CC,A}$,P_{2C}) zum ohmschen Widerstand der 18650 Zelle beitragen.
2. Frequenzbereich II von $5\,\text{kHz}$ bis $10\,\text{Hz}$ (gemessen durch Impedanzspektroskopie): Die Anodenverluste Ladungstransfer ($P_{CT,A}$) und SEI ($P_{SEI,A}$) in Kombination mit der Ionenleitung im Elektrolyten in den Poren dominieren das Impedanzspektrum in diesem Frequenzbereich vollständig. Der Beitrag des Kathodenladungstransfers P_{1C}, der ebenfalls in diesem Frequenzbereich auftritt, ist vernachlässigbar.
3. Frequenzbereich III von $10\,\text{Hz}$ bis $5\,\mu\text{Hz}$ (Zeitbereichsmessungen): In diesem gesamten Frequenzbereich ist die Festkörperdiffusion in der $LiFePO_4$-Kathode ($P_{diff,C}$) der 18650 Zelle der dominierende Verlustprozess. Die Festkörperdiffusion in der Anode ($P_{diff,A}$) hat einen unerwartet kleinen Beitrag.
4. Frequenzbereich IV unterhalb von $5\,\mu\text{Hz}$ (Zeitbereichsmessungen): In der 18650 Zelle wurde hier ein Verlustprozess identifiziert, der nicht in den Experimentalzellen beobachtet werden konnte. Aufgrund dessen und aufgrund seiner extrem hohen Zeitkonstante kann die Homogenisierung, also der Ausgleich der Lithiumkonzentration entlang der Elektrodenwickel, als Ursache dessen angesehen werden.

Bei der vorgestellten Untersuchung der 18650 Zelle sind die frequenzbereichsweise Trennung der Anoden- und Kathodenbeiträge, die Interpretation des ohmschen Widerstandes und die Identifikation der ionischen Leitfähigkeit in den Poren als relevanter Beitrag zum Impedanzspektrum hervorzuheben. Eine vergleichbare Analyse wurde in der Literatur bisher nicht veröffentlicht. Für das korrekte Verständnis der Impedanz kommerzieller Zellen sind diese Erkenntnisse jedoch unverzichtbar.

Contents

1 Introduction

There has been a large increase in attention to electrochemical energy storage systems due to their high potential in future scenarios of energy supply. The foreseeable shortage of primary energy sources and the climate change caused by extensive carbon dioxide emissions have given rise to a search for alternative energy sources. A part of the solution is provided by the exploitation of renewable energies; in Germany this is driven by the so-called *Energiewende*. However, these new ways of generating energy have some disadvantages concerning a reliable and continuous supply because of weather-dependent variations such as wind force or sunshine hours. These variations require a stabilization of the power grid in order to provide a weather-independent power supply.

Another way to reduce the consumption of primary energy sources and the emission of carbon dioxide is an extensive introduction of electric mobility and an increasing degree of hybridization for standard vehicles. Furthermore, this allows for a reduction of local emissions, another big issue due to rising air pollution in the up-and-coming *megacities* where a growing portion of the human population lives.

All of these trends necessitate efficient storage and supply of electric energy. One candidate system for this is the lithium-ion battery, which offers the advantages of a relatively high energy- and power-density and high efficiency [2]. Today, lithium-ion cells are the number-one energy storage technology for consumer electronics such as laptops, mobile phones and power tools. Therefore, a firm groundwork of advanced lithium-ion cell technology exists.

However, electric mobility in particular requires new and stricter properties for lithium-ion cells compared to the current applications. It is necessary to provide very high power density for a good acceleration and recuperation in hybrid cars, whereas pure electric vehicles need a high energy density in order to expand achievable cruising ranges. Moreover, it is essential to improve and to understand the aging behavior of lithium-ion cells as this has, until now, been an unpredictable liability risk for car manufacturers introducing electric mobility extensively.

A further improvement of the performance and the aging behavior of lithium-ion cells requires a comprehensive understanding of the internal mechanisms responsible for performance limits or degradation. An analysis of the cells and the corresponding electrodes by advanced electrochemical measurement and evaluation methods is therefore an inevitable step on the way to next generation batteries for electric mobility.

Goals of this Thesis

The main goals of this thesis are the identification and the quantification of the electro-chemical loss mechanisms occurring in lithium-ion cells and their electrodes. This is achieved by combining electrochemical impedance spectroscopy (EIS) and time domain measurements in order to investigate a wide frequency range from $1\,\mu\mathrm{Hz}$ to $100\,\mathrm{kHz}$. The obtained measurement data are analyzed by the Distribution of Relaxation Times (DRT) which allows for a superior identification of single loss mechanisms. This knowledge is then applied for the design of physically based impedance models that enable a quantification of the polarization of each loss process. The findings are applied to improve the electrode performance systematically, and – for the electrodes extracted from commercially available cells – transferred to the full lithium-ion cell.
Different experimental cell setups are tested in this thesis for the investigation of lab-scale electrodes and electrodes from commercially available cells with the previously described methods. The development of reliable cell configurations was necessary in order to exclude the influence of the measurement setup on the experimental results.

Outline

Chapter 2 gives an introduction to the fundamentals of lithium-ion batteries, electro-chemical measurement techniques and impedance modelling, necessary to understand this thesis. In Chapter 3, an overview of state-of-the-art impedance analysis of lithium-ion batteries and electrodes is presented in order to contextualize the present thesis within the current research. Chapter 4 presents the experimental basics of this study such as experimental cell housings, cell configurations and measurement setup.
The following chapters represent the main results of this thesis:

- Chapter 5 demonstrates the identification of electrochemical loss processes for lab-scale $LiFePO_4$-cathodes via EIS and DRT, followed by the development of a physically based impedance model.

- In Chapter 6, the effect of microstructure variation on the loss mechanisms of lab-scale $LiFePO_4$-cathodes is introduced. Then, the effect of the latter on the discharge performance of these electrodes is analyzed and, finally, an improved cathode structure results.

- In Chapter 7, graphite-anodes extracted from commercially available cells are analyzed by the methods previously presented. A physically based impedance model emerges which is able to represent the loss processes and their interaction with the anode microstructure properly.

- In Chapter 8, a commercially available 18650 cell is investigated by opening the cell and measuring its electrodes in experimental cell configuration. The EIS measurements are complemented by time domain measurements in order to expand the frequency range. The results are subsequently transferred to the 18650 cell leading to a physical interpretation of its relevant loss mechanisms.

Chapter 9 gives a brief summary of the results produced in this thesis.

2 Fundamentals

This chapter gives an overview of the basic knowledge needed to understand the following chapters. It starts with an introduction of lithium-ion cells and their main components and continues with the description of working principle and key properties. The principles of several measurement techniques covering frequency and time domain methods are presented afterwards. Finally, a survey on impedance modelling is given by introducing different equivalent circuit elements and equivalent circuit structures.

2.1 Lithium-Ion Cells

Lithium-ion cells are galvanic elements which are able to convert chemical energy into electrical energy directly by means of electrochemical oxidation and reduction reaction. Furthermore, they are among reversible batteries (secondary cells) where the redox reaction occurs reversibly. This distinguishes them from other types of galvanic cells such as primary cells, in which the redox reaction occurs only in one direction, and fuel cells, in which the reactants have to be supplied continuously. In the literature, there are two terms which are misleading as they are used synonymously: Lithium-ion cell and lithium-ion battery. Originally, a cell is defined as one separate galvanic element, whereas a battery consists of one or several galvanic cells which are connected in a series or in parallel to each other [3]. In this thesis, the term *battery* will be used as the general description of lithium-ion batteries and their properties. The term *lithium-ion cell* will be used for the description of their components and measurements, as all measurements in this thesis were conducted using single cells either in commercial or in experimental cell housings.

2.1.1 Components and Working Principle of Lithium-Ion Batteries

Lithium-ion cells consist of three main components allowing for reversible energy storage. Two electrodes function as host material for lithium-ions. The negative electrode, defined as an anode, is oxidized during the discharge reaction, whereas the positive electrode, defined as a cathode, is reduced during the discharge reaction. Usually, lithium-ions are stored in the active material of these electrodes by intercalation without changing the general lattice structure. The respective partial reaction taking place on the cathode side is shown in Equation 2.1 in the case of a $LiCoO_2$ cathode.

$$Li_{1-x}CoO_2 + xLi^+ + xe^- \underset{charge}{\overset{discharge}{\rightleftharpoons}} LiCoO_2 \tag{2.1}$$

The anode reaction is given by Equation 2.2 in the case of a graphite-anode.

$$LiC_6 \underset{charge}{\overset{discharge}{\rightleftharpoons}} Li_{1-x}C_6 + xLi^+ + xe^- \tag{2.2}$$

Between the electrodes, a lithium-ion conducting electrolyte guarantees the ion transport that enables the spatially divided partial reactions at both reaction sites, as introduced above. The standard electrolyte in lithium-ion batteries is a liquid mixture of solvent and conducting salt. It is therefore necessary to add a fourth, inactive component – the separator – between the electrodes in order to soak up the electrolyte and to separate the electrodes mechanically and electrically.

Assembled in a cell stack, these components work as a reversible energy storage device. Bringing them together leads to a spontaneous reaction which results in a specific Open Circuit Voltage (OCV) of the cell in the case of no external electronic path. It can be calculated using Gibbs free energy ΔG by

$$U_{OCV} = -\frac{\Delta G}{nF} \tag{2.3}$$

where n is the number of charge carriers and F the Faraday constant. ΔG and therefore also the OCV depend on the reaction which takes place. The latter is specified by the active material which is used because each active material provides a characteristic electrochemical potential. The difference of cathode and anode potential defines the specific OCV. The dependence of cathode and anode potential on the lithium concentration in their active material leads to a correlation between OCV and lithium concentration inside the electrodes as well. Each electrode material shows a different characteristic lithium concentration dependence which will be shown in Section 2.1.2.

Connecting an external load to the cell leads to an external electrical current and an internal lithium-ion current. During the discharge process, electrons flow from the anode to the cathode through the external load. In parallel, lithium-ions are transported through the electrolyte from the anode to the cathode side, leading to a changing lithium concentration in the electrodes. This leads to a decreasing OCV, depending on the amount of transported lithium and the active materials used.

The transport of lithium-ions through the cell during discharge comprises of several steps (see Figure 2.1):

1. Lithium must be transported inside the anode active material to the surface which is covered by electrolyte.
2. At this interface, the lithium atom releases an electron and leaves the anode active material. It is dissolved in the electrolyte as lithium-ion and covered by a solvation shell.

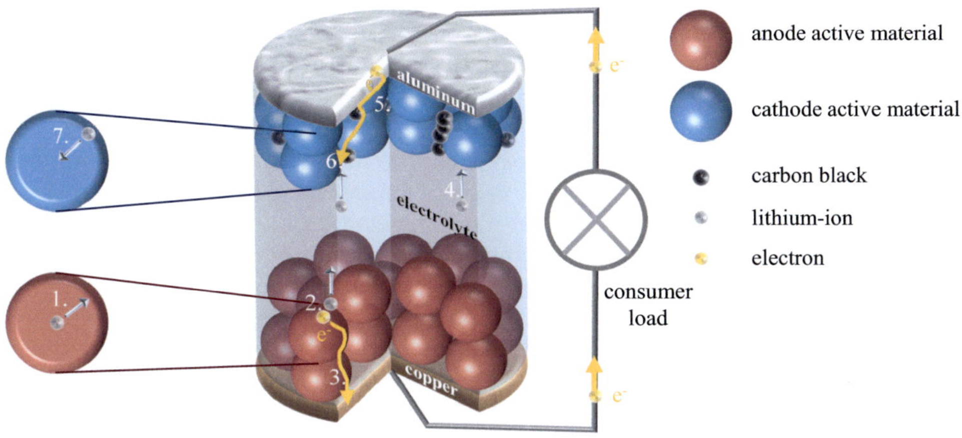

Figure 2.1: Working principle of lithium-ion cells during discharge.

3. The released electron needs to be transported through the anode to the current collector.
4. In parallel, the lithium-ion is transported through the electrolyte to the cathode side.
5. An electron needs to be transported from the cathode current collector to the cathode/electrolyte interface.
6. There, the lithium-ion is inserted into the cathode active material. This insertion process comprises of removing the solvation shell, entering the cathode active material and accepting an electron from the host structure.
7. Then, the lithium has to be transported from the surface into the bulk material in order to equalize the lithium concentration gradients in the active material.

Each of these processes causes a voltage drop denominated as *overpotential*. A detailed explanation of all occurring loss processes will be given Section 2.1.4. All these processes occur in the opposite order for the charge process. Hereby, the application of a charge voltage or charge current leads to a lithium transport from the cathode to the anode side, restoring the original lithium concentrations and increasing the OCV according to the characteristic OCV curve.

One unique aspect of lithium-ion batteries compared to alternative battery technologies is the intercalation of lithium without a significant structural change of electrodes, instead of an electrochemical reaction with the electrodes' material and the electrolyte. This increases their cycling efficiency and cycling stability. Nevertheless, the insertion of lithium causes mechanical stress for the electrodes because, depending on the active material, volume changes occur.

2.1.2 Structures and Materials

The components of lithium-ion cells must meet several criteria to be suitable for application. A large variety of candidate materials is available for the electrodes as well as for the electrolyte because until now, no perfect material combination was identified. For the electrolyte it is important to provide a very high lithium-ion and a minimum electronic conductivity. Furthermore, an optimal electrolyte would be electrochemically stable at the electrodes' potential and mechanically stable for safety reasons. An inexpensive and non-toxic electrolyte would also be desirable.

For electrodes, the following properties are required:

- high specific lithium storage capacity
- high ionic and electronic conductivity
- low volume change during lithium intercalation
- long cycle life
- large difference between electrochemical potential of cathode and anode
- high safety during operation and stability against electrolyte
- easy available and inexpensive

As the pure active material (material for lithium storage) does not provide these properties satisfactorily, all electrodes are fabricated as porous composite layers from several components (as sketched in Figure 2.1):

- current collector: gives mechanical stability and provides the electrons
- active material: stores the lithium
- carbon black: increases the electronic layer conductivity
- binder: keeps the electrode components together

The porosity allows for an enhancement of the electrochemically active surface area and a shortening of the lithium transport paths inside the active material. The fabrication of porous electrode layers by mixing the materials mentioned above allows for a targeted tuning of the desired electrode parameters. It is furthermore crucial to choose an optimum layer thickness, porosity, particle size and material composition. It is possible to adapt these parameters according to the requirement to achieve either high energy or high power density.

Besides mixing the active material with these components, it is additionally possible to mix different active materials in one electrode layer. These so-called blend electrodes combine the advantages of different active materials in one electrode [4, 5]. This allows for the development of safer electrodes with increasing energy and power density and longer cycle life.

There are three groups of active material, which can distinguished by their lattice structure (shown in Figure 2.2). The lattice structure has an impact on the diffusion paths of lithium inside the active material as 1D-, 2D- and 3D-diffusion paths may occur [6]. The transport and intercalation of lithium is improved when more free paths are available. Moreover, the thermal stability of these materials and their stability against over-charge and over-discharge is influenced by their structure.

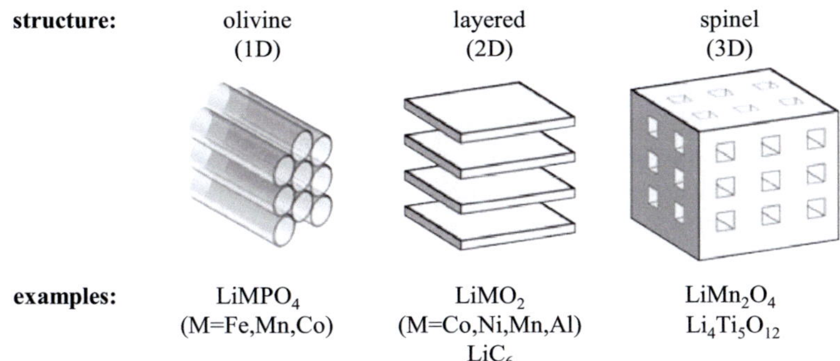

Figure 2.2: Different electrode materials, their corresponding lattice structure and the resulting lithium diffusion paths [4, 6, 7].

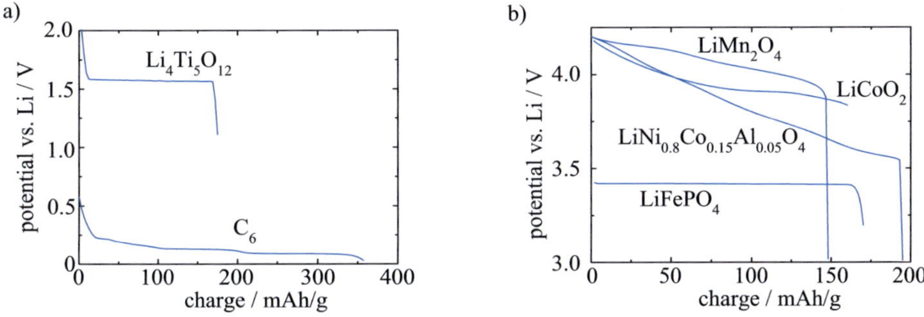

Figure 2.3: Open Circuit Potential (OCP) for common a) anode and b) cathode materials.

Another important issue for the choice of the active material is its electrochemical potential versus lithium. The cathode potential should be as high as possible whereas the anode potential should be relatively low. Furthermore, the potential of each electrode shows a characteristic dependency on lithium concentration, as presented in Figure 2.3. Depending on the application, a relatively constant or a continuously changing potential can be desirable. The following two sections will introduce different electrode active materials and their properties.

2.1.2.1 Cathodes

In general, layered metal oxides (LiMO$_2$), spinel oxides (LiM$_2$O$_4$) and phospho-olivines (LiMPO$_4$) are suitable as cathode materials in lithium-ion batteries [8]. All of these materials have a relatively low electronic conductivity which makes the fabrication of optimized composite layers necessary. Therefore, highly conductive additives such as carbon black or vapor grown nano fibers (VGCF) are added and the active material surface is usually coated by a thin carbon layer in order to increase its intrinsic electronic conductivity. Moreover, it can be advantageous to fabricate complex microstructures with small active material particles and larger agglomerates in order to increase the energy density without loosing the performance at high currents.

These composite cathode layers are coated on an at least 15 μm thick aluminum foil as current collector [9]. Aluminum is used as it is electrochemically stable at the standard cathode potentials.

LiCoO$_2$　　One of the first cathode materials commercialized by Sony was LiCoO$_2$, which was introduced by the Goodenough group in 1980 [10, 11]. It is a layered oxide providing a relatively high average electrode potential (3.9 V, see Figure 2.3) and a high specific capacity of 150 mAh/g [9]. However, nowadays two issues prevent the use of LiCoO$_2$ as cathode material. First, the price of cobalt increases steadily and makes it economically less attractive. Second, it is not safe for charging above 4.4 V and for high temperatures. In both cases, the layered structure becomes unstable and causes an exothermic reaction which is a large safety risk.

LiNi$_x$Mn$_y$Co$_{1-x-y}$O$_2$ (NMC)　　An improvement of layered oxides was achieved by combining several metals in one metal oxide with different stoichiometries according to LiNi$_x$Mn$_y$Co$_{1-x-y}$O$_2$ [12, 13]. These structures provide higher stability than LiCoO$_2$-cathodes and combine the major advantages of each material, namely high capacity (LiCoO$_2$), fast discharge (LiNiO$_2$), stability for over-charge and low cost (LiMnO$_2$). A specific capacity of 130 mAh/g to 160 mAh/g and a high average discharge potential of 3.9 V can be achieved, depending on the fraction of each component [9]. A very detailed discussion of different compositions of NMC and other cathode materials can be found in Refs. [14, 15].

LiNi$_x$Co$_y$Al$_{1-x-y}$O$_2$ (NCA)　　In 2001, NCA was proposed as a new cathode material in the class of layered oxides [16]. It provides a high specific capacity up to 200 mAh/g and a good cycle life [5, 17, 18]. However, the drawbacks of NCA are high irreversible capacity loss during the first cycle [19] and poor thermal stability [20]. Therefore, it has recently been investigated as material for blend electrodes in combination with LiMn$_2$O$_4$ [4, 5].

LiMn$_2$O$_4$　　LiMn$_2$O$_4$ spinel has drawn attention as an alternative electrode material because it is nontoxic, inexpensive and shows a high thermal stability [21–23]. However, it has a low specific capacity of around 120 mAh/g [9] and poor cycle life. Nevertheless, it has recently been used in blend electrodes as it can increase stability and rate capability and its disadvantages can be compensated for by using a second component such as NCA [4, 5].

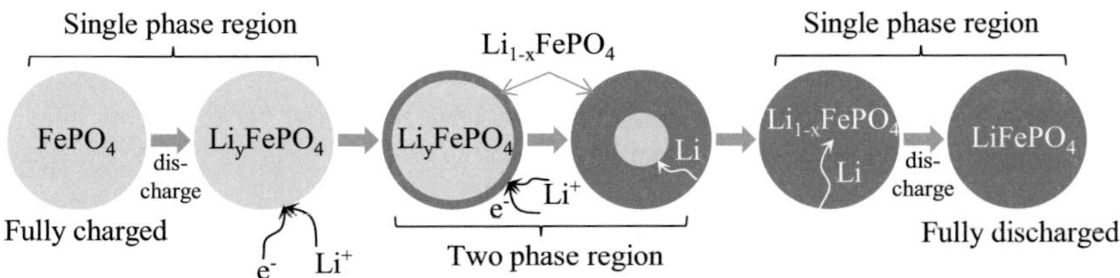

Figure 2.4: Shrinking-core model for LiFePO$_4$ according to Refs. [31, 32].

LiFePO$_4$ LiFePO$_4$ is the cathode material investigated in this thesis. It was introduced by Goodenough et al. in 1997 as a cathode candidate material for lithium-ion cells [24, 25]. In contrast to the other cathode materials introduced, LiFePO$_4$ is a phospho-olivine and has therefore an olivine type structure. This leads to the main advantage of LiFePO$_4$. The olivine structure is very stable and provides a high level of safety as it does not show an exothermic reaction for over-discharge or for increasing temperatures. Furthermore, it is inexpensive, environmentally benign, abundant [26] and shows a good cycle life as the volume changes during lithium insertion are relatively small (6.8 %) [9]. Its practical specific capacity is 160 mAh/g, which is a good value compared to the other cathodes. The disadvantages of LiFePO$_4$ are first the low potential of 3.43 V, leading to a low energy density, and second its extremely low ionic and electronic conductivity [27, 28]. The low ionic conductivity is related to the olivine structure as that provides only one-dimensional diffusion paths (presented in Figure 2.2), leading to a limited lithium transport inside the active material.

In order to overcome the low ionic conductivity, the synthesis of LiFePO$_4$ nano-particles was successfully studied in many groups and resulted in high rate capability for LiFePO$_4$ particles due to short diffusion paths [26]. Moreover, an increase of electronic conductivity was achieved by coating the LiFePO$_4$ particles and adding multi-walled carbon nanotubes (MWCNTs) [26, 29]. Furthermore, vapor-grown carbon fibers (VGCF) are introduced to increase the electronic layer conductivity [30]. Obviously, the development of complex microstructures played an important role to make LiFePO$_4$ applicable for use in lithium-ion cells.

One advantage of LiFePO$_4$ is its almost constant potential of 3.43 V. This plateau is caused by a two-phase behavior, already found by Padhi et al. in 1997 [24]. LiFePO$_4$ forms a lithium-rich (LiFePO$_4$) and a lithium-poor (FePO$_4$) phase, separated by a phase boundary (Figure 2.4). A simplified model to describe this behavior was developed by Newman et al. [31, 32]. The model known as *shrinking-core model* assumes a radial movement of the phase boundary from particle surface to particle center. This would result in a lithium diffusion through the phase in the outer shell to the inner shell and a movement of the phase boundary according to the SOC. The olivine structure of LiFePO$_4$ further allows the model assumption of one-dimensional diffusion, leading to a straight phase separation perpendicular to the diffusion paths [33] and an adaption of the shrinking core model to describe a one-dimensional phase separation.

2.1.2.2 Anodes

Lithium metal would be the best anode material with regard to the basic requirements of anodes for lithium-ion cells. It has the lowest possible potential, the highest specific capacity as no host material is necessary and high electronic conductivity. However, there are several drawbacks which make lithium not applicable in rechargeable lithium-ion cells. Lithium metal is chemically very active. This leads not only to safety problems in case of an accident, it also causes strong reactions with the organic electrolyte inside the cell. During cycling, this leads to the consumption of active lithium as the lithium reacts after its redeposition with the electrolyte. This reaction results in a protective layer called Solid Electrolyte Interphase (SEI) layer. During the deposition and stripping of lithium, the SEI layer is broken up, a new surface film is formed and lithium is consumed permanently [34]. Due to irregular deposition of lithium on the anode surface it is also possible that fiber-like lithium dendrites occur. If these dendrites are isolated from the lithium anode itself, the lithium is irreversibly lost. If they are not isolated but grow steadily, this may result in electrical shorting between the electrodes and a destruction of the entire cell. This happened in 1989 to batteries of Moli Energy and led to the recall of a high number of lithium-ion cells [9]. This safety risk is one of the main reasons why lithium metal is not applied in rechargeable lithium-ion cells any more.

However, lithium metal has several advantages for the investigation of electrode materials in experimental cells and is therefore applied as the counter electrode in lithium-ion battery research. It provides a surplus amount of active lithium and therefore does not limit the cell capacity. Moreover, no lithium diffusion occurs as the lithium is directly deposited on the anode surface. This makes lithium metal a widespread counter electrode for the use in experimental cells. Furthermore, primary lithium-ion cells with lithium metal anode are now available [34]. The research on next generation anodes which are very similar to lithium metal anodes focuses on lithium alloys (lithium-aluminum, lithium-silicon or lithium-tin) [35]. They reduce some problems of lithium metal at the expense of slightly smaller energy densities due to additional weight and increasing anode potential.

Today, mainly carbon-based anodes are applied and, in rare cases, $LiTi_5O_{12}$ is used as the anode active material. Both kinds of anodes are coated as a porous layer on copper foil as a current collector. The thickness of typical anode layers varies between $20\,\mu m$ and $100\,\mu m$, depending on the application in high power or high energy cells. The current collector is a copper foil, usually around $10\,\mu m$ thick.

Carbon based Anodes   The first application of carbon as an anode material in a commercially available lithium-ion cell was introduced by Sony in 1991 [36]. There are two classes of carbon anodes, graphitic and nongraphitic. Nongraphitic carbon - also called amorphous carbon - consists of amorphous areas and graphitic structure elements, partially crosslinked [37,38]. However, it does not show a far-reaching crystallographic order. It is able to insert 0.5 - 0.6 lithium atoms per C_6 unit and therefore offers a specific capacity of $200\,mAh/g$. The voltage varies steadily between $700\,mV$ and $50\,mV$

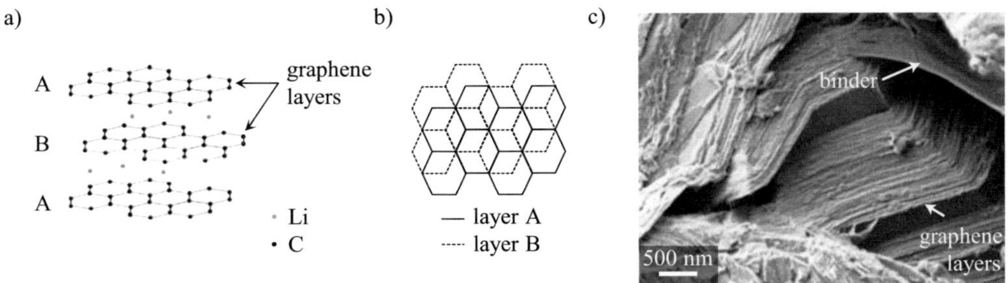

Figure 2.5: a) Schematic drawing of the crystal structure and stacking of graphene layers for hexagonal graphite electrodes. b) Perpendicular view on the Basal planes of hexagonal graphite. Modified and redrawn from Ref. [6]. c) SEM-image of real graphite structure by Moses Ender, IWE.

versus lithium metal, depending on the lithium content [9].

The graphitic structures are more popular among the carbon anodes. There are different kinds of graphitic carbons, namely *meso carbon micro beads* (MCMB), *natural graphite, synthetic graphite* and, as a part of the synthetic graphites, *highly oriented pyrolytic graphite (HOPG)*. These graphitic materials are produced differently and vary in particle size, particle morphology and the amount of defects in their crystal order [37]. The carbon atoms are arranged in a planar hexagonal network forming so called graphene layers. Those are kept together and aligned by Van der Waals forces in AB-order to form graphite (see Figure 2.5), which is able to store lithium reversibly between its layers.

Graphite-anodes are able to store up to one lithium atom per C_6 unit and therefore, to achieve a theoretical specific capacity of 372 mAh/g. During this lithium intercalation, the layer distance and therefore the graphite volume increases by about 10.3 % and changes its order from AB to AA. Consequently, the carbon atoms directly face each other between two neighboring graphene layers in the lithiated state [37]. The electrochemical potential of graphite-anodes versus lithium metal is lower than for nongraphitic anodes and varies for most practical lithium concentrations between 250 mV and 50 mV. The potential does not correlate linearly to the lithium concentration. This is connected to a staging mechanism during lithiation of the graphite.

Graphite is known to intercalate lithium-ions in a regular order between the graphene layers [37, 39]. This effect is known as stage formation or staging. This thermodynamic phenomenon is related to the energy required to open the gap between the graphene layers for lithium intercalation. The stage number corresponds to the number of layers which are located between the two nearest-guest layers [37]. Accordingly, stage I is the lithium concentration where lithium is intercalated between all graphene layers, corresponding to LiC_6. The next stage corresponds to the half lithium concentration of LiC_{12}. Furthermore, stage III and IV were found to form correspondingly for lower lithium content [37]. However, the exact formation of those stages above stage II is not

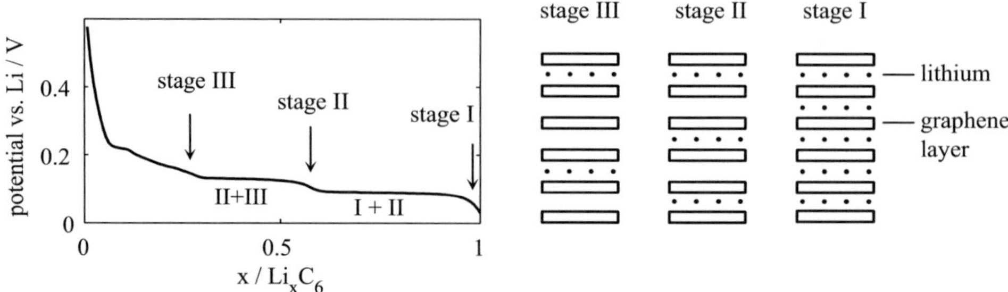

Figure 2.6: OCP and corresponding stages for a graphite-anode.

finally understood [40].

The effect of this characteristic intercalation behavior on the electrochemical potential of graphite-anodes is shown in Figure 2.6. At each stage, a change of the potential occurs whereas between two stages, a mixture of two phases occurs and causes a voltage plateau. The potential of graphite-anodes is below 0.8 V for almost all lithium concentrations. This is below the electrochemical stability window of standard liquid electrolytes and therefore causes the formation of a SEI. The details about SEI formation and its components will be explained in Section 2.1.2.3. The SEI layer has two important effects on graphite-anodes which are indispensable to using graphite as anode material. First, it passivates the graphite surface and prevents a further decomposition of the electrolyte. Without SEI, the electrolyte would be decomposed continuously and lead to a serious aging of the electrolyte. Second, the SEI layer is able to prevent the intercalation of solvated lithium-ions into the graphite-anodes. Otherwise, such intercalation would lead to an exfoliation of graphene layers and lead to a loss of active material [41].

Li$_4$Ti$_5$O$_{12}$ Li$_4$Ti$_5$O$_{12}$ is the only metal oxide which is used as anode material. Its main advantage and disadvantage is its high potential plateau at 1.55 V versus Li/Li$^+$. Therefore, no SEI layer is formed as the electrolyte is stable at these potentials. This fact, together with almost zero volume change during cycling (*zero strain material*), makes Li$_4$Ti$_5$O$_{12}$ have the best cycling stability among other anode materials [9]. However, the high potential also reduces the cell voltage and, with that, the achievable energy densities. Li$_4$Ti$_5$O$_{12}$ itself offers a practical specific capacity of 150 mAh/g, which is drastically less than graphite-anodes.

2.1.2.3 Electrolytes

The electrolyte functions as a lithium-ion conductor between the electrodes. Until now, usually non-aqueous liquid electrolytes have been used in lithium-ion batteries, containing three main components:

- lithium-salt for ionic conductivity (i.e. $LiPF_6$,$LiClO_4$,$LiBF_4$,$LiAsF_4$)
- polar solvent component for lithium-salt solubility (i.e. ethylene carbonate (EC) or propylene carbonate (PC))
- non-polar solvent to decrease the viscosity (i.e. ethyl methyl carbonate (EMC), dimethyl carbonate (DMC) or diethyl carbonate (DEC))

Mixtures with more than two kinds of solvent are possible as well [42]. The proportion of components influences the electrolyte properties, for instance conductivity, viscosity and stability [42]. Moreover, the lithium charge transfer between electrolyte and electrodes is affected [43, 44]. One standard electrolyte mixture is a 1M (1 mol/l) solution of $LiPF_6$ in a mix of (i) EC:EMC or (ii) EC:DMC (both 1:1), providing conductivities of (i) 10.7 mS/cm and (ii) around 8 mS/cm at 20 °C [45]. However, this is only a basic electrolyte as there are many additives that can improve its properties. Vinylene carbonate (VC) is one of the most popular additives and was also used later in this thesis. According to studies in the literature it is able to improve the formation of the reaction layer on the anode surface and to reduce the amount of irreversible lithium loss [42, 46, 47].
The electrochemical window in which the liquid electrolytes are stable should be as large as possible (0 V to 5 V). However, they are usually not stable below 0.8 V and above 4.5 V, which causes a decomposition of the electrolyte at the electrode surface. This results in a reaction layer denominated as SEI. A uniform SEI formation is as least as important as a good ionic conductivity in order to limit the amount of electrolyte decomposed and the following rise of interface resistance. Therefore, the electrolyte composition of commercially available lithium-ion cells is one of the most closely guarded secrets of cell manufacturers.
New electrolyte concepts are under investigation for future lithium-ion batteries. These cover the use of polymer electrolytes, solid electrolytes or ionic liquids. A good overview of these concepts is given in Ref. [8].

Solid Electrolyte Interphase (SEI)   As described above, the formation of a reaction layer at the anode/electrolyte interface (and less pronounced also at the cathode/electrolyte interface) is inevitable for standard anodes in liquid electrolytes. This fact is well-known; however, the reactions causing this layer and its properties are not well understood. It is formed from the decomposed liquid electrolyte and thus contains parts of the solvent and the conducting salt including lithium. It is assumed that the SEI consists of several layers or micro-phases composed of different components such as $LiCO_2-R$, LiF, Li_2CO_3, Li_2CO_3, Li_2O, lithium alkoxides or nonconductive polymers. The exact composition depends on the electrolyte and on the anode material used. Some SEI models propose a thin and compact layer at the anode surface made from inorganic components and a second layer which is more porous and made from organic components on the electrolyte side. Further details can be found in Refs. [41, 48].
The SEI should have the following properties for the application of lithium-ion cells: First, it should be thin and provide a good lithium-ion conductivity in order to minimize its influence on the interface resistance. Second, the SEI-formation should be limited

to the first cycles in order to prevent an ongoing aging of the electrolyte, a continuous lithium consumption causing a capacity fade and an increasing interface resistance. A reduced SEI formation is achieved if the SEI layer is electronically nonconducting. It should furthermore be mechanically and thermally stable as a repeated break-up and new formation of the SEI during operation would also lead to an aging of the electrolyte. The optimization of the formation process is a proprietary knowledge of battery producers and is not accessible.

A quantification of the SEI contribution to the internal loss mechanisms of lithium-ion cells is an important step in order to obtain an understanding of its effect on the cell performance and on cell degradation. This is a crucial step as the SEI growth is known to be an important factor for a limited cell performance and a decreasing cell capacity [49, 50]. An unambiguous identification of the SEI contribution would allow for a specific improvement of the electrolyte with direct measurement of its effect on the SEI formation. Furthermore, this would enable a separation of the losses by the charge transfer process and by the transport through the SEI layer. With that, it would be possible to separate a degradation of the electrolyte affecting the charge transfer and a growth of the reaction layer increasing the transport losses through it.

2.1.2.4 Separator

In the case of liquid electrolytes, a separator is needed to separate the electrodes mechanically. Its most important properties are mechanical stability, a lack of electronic conductivity and a minimum influence on the ionic conductivity of the electrolyte. Furthermore, it has to be electrochemically stable in the electrolyte solution and at the contact area to the electrodes. Usually a $20\,\mu m$ to $25\,\mu m$ thick polyethylene (PE) or polypropylene (PP) layer with a homogenous porosity (around $50\,\%$) is used, as it fulfills these requirements satisfactorily. During the last few years, an additional functionality called *shut-down* was developed [9]. There, the core of the separator melts due to heating of the cell in case of internal or external short circuits. It closes the ionic and electronic path between the electrodes and thus prevents a further heating and possible explosion of the cell. This shut-down mechanism is irreversible and allows no further use of the cell.

For experimental lithium-ion cells, which are used in the research and development of lithium-ion batteries, another kind of separator is used. It is made from glass fiber and shows a very high porosity up to $85\,\%$. These separators are far thicker ($200\,\mu m$ to $1.55\,mm$) in order to prevent a short circuit in the experimental cell setup and to provide a large electrolyte reservoir. This is very important for experimental cells as a lot of free volume is available in experimental cell housings. An evaporation of electrolyte can therefore not be prevented and necessitates a large electrolyte reservoir.

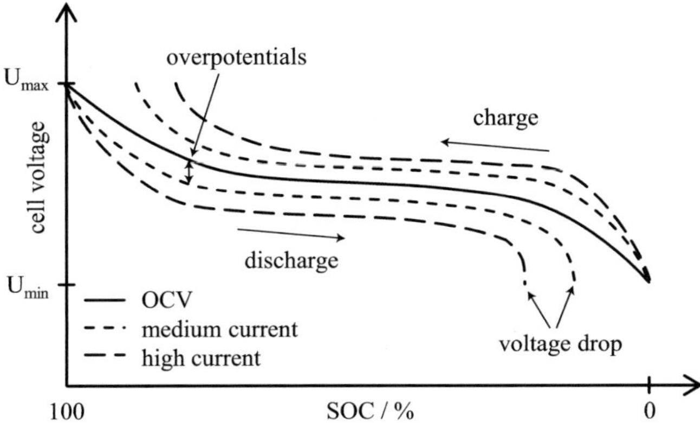

Figure 2.7: Schematic OCV and typical charge and discharge curves for a lithium-ion cell.

2.1.3 Charge and Discharge Curves

The characteristic charge/discharge curve of a lithium-ion cell depends on the active materials used and on the electrochemical loss mechanisms described in the following section. The active materials produce a characteristic, SOC dependent OCV which is defined by the difference of cathode and anode OCP. The detailed correlation between OCV, OCP and electrode capacities is explained in Section 2.1.5.5. The OCV curve shows a high cell voltage for fully charged cells and a decreasing cell voltage for decreasing SOC (see Figure 2.7). The upper and lower cut-off voltage U_{max} and U_{min} are defined by the electrodes used. Furthermore, charge and discharge curves are given for different current rates in Figure 2.7. During discharge, the cell voltage is below the OCV. This gap is caused by the overpotentials and it depends on the discharge current. Depending on the loss process, this current dependency can be linear (i.e. for the ohmic resistance) or non-linear (i.e. charge transfer process, Section 2.1.4.3).

Not only the cell voltage is affected by the discharge rate; the cell capacity is reduced for high discharge rates as well. This is caused by the sharp voltage drop indicated at the end of the discharge process in Figure 2.7, stopping the discharge at U_{min}. The sharp voltage drop can be explained by the overpotentials discussed in the next section. Three fundamental cases can be distinguished, each of them caused by the depletion of one of the following species. Lithium-ions can be depleted in the electrolyte due to limited lithium transport in the porosity. Furthermore, lithium depletion can occur at the surface of the active material if the solid state diffusion is the limiting factor. Lastly, the transport of electrons can also be limited as the electronic conductivity is very small for some electrode material. All of these cases lead to an increasing overpotential at the electrode/electrolyte interface and cause a termination of the discharge procedure. It is important to mention that the cell is not completely discharged at this time. After a certain time of voltage relaxation, the cell can be discharged further until the depletion of one of the species reoccurs at the interface. All of these mechanisms hold true for the charge of lithium-ion cells as well.

The differentiation between which of these mechanisms limits the discharge capacity is ambiguous. It strongly depends on the microstructure of the electrode and can be a mix of these three mechanisms. This makes the optimization of electrode microstructure one of the most deciding factors for a good cell performance next to the choice of the active material. In Section 6, an optimization of $LiFePO_4$-electrodes concerning these limiting mechanisms will be introduced.

2.1.4 Loss Mechanisms

As described above, lithium-ions are transported between the electrodes during charge and discharge of the cell. Furthermore, electrons have to be transported from the current collectors to the position where lithium is intercalated. All of these transport mechanisms cause certain overpotentials which decrease the cell voltage during discharge and increase the latter during charge, as described in the previous section. In the following sections, all possible loss mechanisms are illustrated and described by physical equations. Two general terms are used to classify two groups of loss processes. Those showing pure ohmic behavior will be specified as *ohmic losses*, whereas the remaining ones will be specified as *polarization losses*.

2.1.4.1 Lithium Transport in the Electrolyte

Lithium-salt (see Section 2.1.2.3) allows for the transport of lithium-ions in liquid electrolyte. When the lithium-salt (i.e. $LiPF_6$) is dissolved in the electrolyte, a lithium cation (Li^+) is separated from the corresponding anion (PF_6^-) and surrounded by solvent molecules called a *solvation shell*. These lithium cations and the anions provide an ionic conductivity in the electrolyte. However, for the use in lithium-lion cells only the portion of lithium-ion conductivity plays an important role.

The lithium-ion transport is driven by two forces: Migration, a transport induced by electric field, and diffusion, induced by concentration gradients. These two kinds of transport mechanisms cause a certain dynamic for the lithium transport through the electrolyte. However, this dynamic is very fast and is therefore assumed to be part of the ohmic losses [51].

The relation between ionic conductivity and electrolyte resistance is given by ohm's law.

$$R = \sigma_{ion} \frac{L_{el}}{A_{el}} \tag{2.4}$$

Here, σ_{ion} gives the specific ionic conductivity, L_{el} the thickness and A_{el} the cross section of the electrolyte. For the ohmic resistance in lithium-ion cells it is important to take the microstructure of separator and electrodes into account as well. The electrolyte resistance is thereby increased because the volume of separator and electrode do not contribute to the lithium transport. This can be expressed by a lower effective ionic conductivity

$$\sigma_{ion,eff} = \sigma_{ion} \frac{\epsilon}{\tau_p},$$

(2.5)

using porosity ϵ of the surrounding medium and tortuosity τ_p of the pore volume. The tortuosity gives a geometric factor, how drastically the lithium transport path is extended by the curvature of the pore fraction [52].

The calculation of ohmic resistance using this equation simply includes the effect of geometric parameters. Additional effects such as a complex interaction between lithium-ion conduction and the surrounding material are not considered. This is known to have an effect on gas transport inside an electrodes porosity for fuel cells. However, it is not well understood for the lithium-ion transport in liquid electrolyte.

2.1.4.2 Electron Transport through the Electrodes

The transport of electrons to the reaction sites is a crucial factor for lithium-ion electrodes. It depends on the electronic conductivity of current collectors, the active material and other electronically conducting electrode components. In contrast to commonly used anode materials, most cathode active materials provide a very low electronic conductivity [8]. Therefore, electronically conducting components such as carbon black (CB) are introduced to the electrode structure. Furthermore, the active materials are coated in order to increase their conductivity at the particle surface. For the same reasons it is important to fabricate the electrode layers so that they are homogeneous and well-connected to obtain low interface resistances between active material particles and between electrode layer and current collector. These interfaces are not well-investigated; however, it can be expected that they contribute significantly to the internal loss mechanisms of lithium-ion cells.

2.1.4.3 Interface Loss Processes

Another important step occurs at the interface between electrode and electrolyte where lithium-ions are inserted and the ionic transport passes over to an electronic conduction. This interface process comprises several steps such as the breaking up of the solvation shell, the transfer of an electron to the lithium-ion and the insertion of lithium into the host material (see Figure 2.8). These steps are usually summarized as the *charge transfer* process. Furthermore, the transport through reaction layers at the interface (see Section 2.1.2.2) may produce overpotentials as well.

The charge transfer is often described by the Butler-Volmer equation in order to give a relation between overpotential and current density:

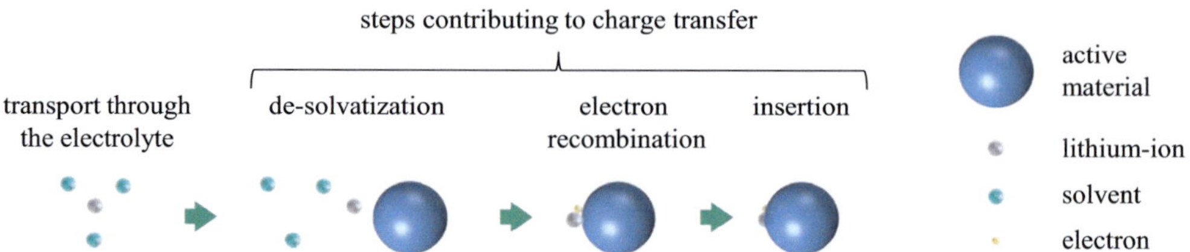

Figure 2.8: Sketch of steps contributing to the charge transfer process at the electrode/electrolyte interface.

$$j = j_{0,el} \left(e^{\alpha \frac{nF\eta_{CT,el}}{R_g T}} - e^{-(1-\alpha)\frac{nF\eta_{CT,el}}{R_g T}} \right) \tag{2.6}$$

Here, $j_{0,el}$ denotes the exchange current density and $\eta_{CT,el}$ the overpotential of the charge transfer process with the number of exchanged electrons n_e (set to one for lithium-ions) and the apparent charge transfer coefficient α. α is a symmetry factor for the reaction which may change for different lithium concentrations. The dependency of charge transfer overpotential on lithium concentration can be seen in an SOC-dependent $j_{0,el}$ and varies for different electrodes. It can be calculated from charge transfer resistance by linearization according to r_{ct} by $j_{0,el} = \frac{R_g T}{nF r_{ct}}$ [53]. Moreover, the charge transfer is facilitated by higher temperatures (for details see Section 2.1.4.5). The use of a Butler-Volmer equation combines all steps of charge transfer in one overpotential. A separation of the individual processes is not possible. However, for a description of occurring overpotentials it is usually sufficiently accurate.

2.1.4.4 Solid State Diffusion

The lithium is to be transported from the surface into the center of the active material particles and vice versa. This process of lithium transport is called solid state diffusion, as it is driven by a concentration gradient. It can be described by Ficks' first and second law and occurs along one-dimensional, two-dimensional or three-dimensional diffusion paths, according to the lattice structure of the active material. The concentration difference between lithium concentration at the surface and equilibrium concentration causes the diffusion overpotential. The relation between concentration difference and overpotential is usually described by the Nernst equation [54]. The diffusion overpotential depends on the applied current, the structure of the active material and the temperature. Moreover, it depends on the length of the transport path and therefore on the particle size. Smaller active material particles reduce the diffusion path length and therefore reduce the diffusion overpotential.

2.1.4.5 Temperature Dependency of Loss Mechanisms

The temperature dependency of electrochemical mechanisms such as ionic conductivity or charge transfer can be described by Arrhenius' law:

$$k = e^{-\frac{E_{act}}{R_g T}} \tag{2.7}$$

Here, k denotes the rate constant of the underlying electrochemical process, E_{act} the activation energy, R_g the universal gas constant and T the absolute temperature in Kelvin. The rate constant k is usually reciprocally proportional to the polarization resistance of the underlying loss process, enabling the determination of E_{act} by temperature dependent measurements of the polarization resistance.

2.1.5 General Terms and Definitions

2.1.5.1 Specific Capacity

The theoretical lithium capacity of an electrode material for lithium-ion batteries is denominated as specific capacity. It is defined for each active material by Equation 2.8 without taking any other cell component into account.

$$C_{spec,M} = \frac{n \cdot F}{M_{molar}} \qquad C_{spec,V} = \frac{n \cdot F}{V_{molar}} \tag{2.8}$$

Here, n is the number of electrons per reaction, F the Faraday constant and M_{molar}/V_{molar} the molar mass/volume of the active material itself. The specific capacity gives the amount of charge which can be stored in a certain mass or volume of active material. For several active materials, the lithium cannot be extracted completely. Therefore, the value for n is not always set to one (i.e. $n = 0.6$ for $LiCoO_2$).

2.1.5.2 Energy and Power Density

Another important practical value is given by energy and power density of lithium-ion cells. For the first one, a theoretical value can be calculated from the specific capacity by

$$w_{th,M} = U_{th} \cdot C_{spec,M} = \frac{U_{th} \cdot n \cdot F}{M_{molar}} \qquad w_{th,V} = U_{th} \cdot C_{spec,V} = \frac{U_{th} \cdot n \cdot F}{V_{molar}} \tag{2.9}$$

This value is not unambiguous as the theoretical voltage U_{th} is not constant. Almost every electrode potential depends on the SOC. The practical energy density is more meaningful for the practical application. It can be calculated by discharging the cell and using Equation 2.10:

$$w_{pr,m} = \frac{\int_{t_0}^{t_{end}} \left(U_{cell} \cdot I_{discharge}\right) d\tau}{m_{cell}} = \frac{\overline{U}_{cell} \cdot I_{discharge} \cdot \left(t_{end} - t_0\right)}{m_{cell}}. \tag{2.10}$$

It can also be related to the cell volume in order to calculate the volumetric energy density. The energy density depends on the discharge current $I_{discharge}$ as higher currents lead to higher overpotential. This was already demonstrated for the discharge curves in Figure 2.7. Another measure available to evaluate the performance of electrodes or cells is the power density. It is determined by discharging the cell and using Equation 2.11:

$$p_{pr,m} = \frac{\int_{t_0}^{t_{end}} \left(U_{cell} \cdot I_{discharge}\right) d\tau}{\left(t_{end} - t_0\right) \cdot m_{cell}} = \frac{\overline{U}_{cell} \cdot I_{discharge} \cdot \left(t_{end} - t_0\right)}{\left(t_{end} - t_0\right) \cdot m_{cell}}. \tag{2.11}$$

This measure is even more dependent on the discharge parameters than the energy density. Not only the discharge current $I_{discharge}$ but also the length of the discharge procedure is a crucial parameter. A very short discharge pulse at high SOC delivers a high power density for a short time, whereas a long discharge until low SOC and low cell voltage delivers a lower power density for a longer time. Therefore it is important to standardize these parameters in order to obtain useful comparisons. In this thesis, only complete discharge processes between upper and lower cut-off voltage after a defined charge procedure will be used as a standard to determine energy and power densities. A good visualization to compare different electrodes or cells by their energy and power density is provided by the Ragone plot in Figure 2.9. It plots energy density versus power density for different cells or discharge currents. Figure 2.9 compares the general difference between high energy and high power cells for different discharge rates. The former provides a higher energy density for low discharge currents. However, the latter is able to provide high power densities which cannot be achieved by the high energy cell due to its large internal resistance. These different characteristics can be achieved by optimizing the cell and its components for different application. The high energy cell is thereby optimized to use a minimum of inactive material, whereas the high power cell is optimized to have a very small internal resistance at the cost of additional inactive material. These optimizations for different applications are done on the cell as well as on the electrode level. The latter will be discussed for lab-scale $LiFePO_4$-cathodes in Section 6.

In practical application it is common that the cell specifications limit the discharge rates in the way that the characteristic drop of energy density for high discharge rates cannot be observed as this would happen out of the specified parameters.

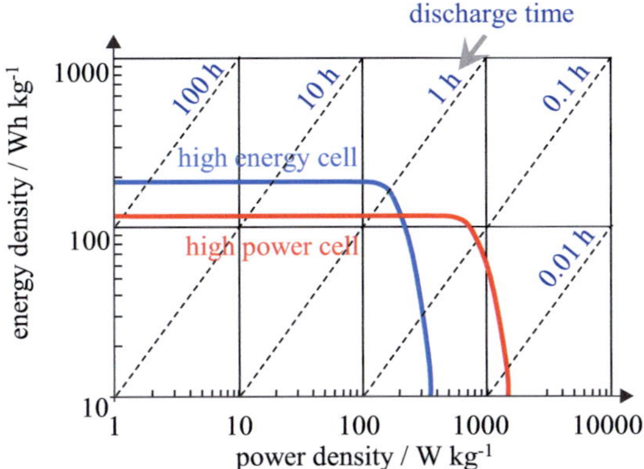

Figure 2.9: Characteristic of high energy and high power cell in a Ragone plot. Modified and redrawn from Ref. [55].

2.1.5.3 C-rate

The C-rate gives a relation between discharge current I and cell capacity C_{cell} in Ah by

$$I_{xC} = x \cdot C_{cell}/1h. \tag{2.12}$$

x takes any positive value. For instance, I_{1C} is given by the nominal cell capacity [1] in Ah divided by one hour. For a 5 Ah-cell this would mean that $I_{1C} = 5A$, $I_{10C} = 50A$ and $I_{\frac{C}{10}} = 0.5$ A. This definition allows for a better comparability between cells with different capacities.

2.1.5.4 State of Charge

There are many ways to define the State Of Charge (SOC) of a lithium-ion cell. In this thesis, it is defined by the discharged charge in relation to the real cell capacity C_{cell}. Setting an SOC therefore requires the following steps:

1. Charge the cell using the standard CCCV-charge protocol (CCCV: Constant Current/Constant Voltage). This procedure comprises constant current phase until the upper cut-off voltage is reached and a subsequent constant voltage phase. The CCCV is finished when the charge current falls below C/10.
2. Discharge the cell with nominal discharge current (1C in this thesis) until the lower cut-off voltage is reached. C_{cell} is then given by the amount of charge.

---

[1] Capacity given by the cell manufacturer. It can be achieved by discharging the fully charged cell under nominal conditions (nominal current, nominal temperature, given cut-off voltages)

3. Afterwards, setting the SOC starts with charging the cell (congruent to step 1) and concludes with discharging the specific amount of charge until the desired SOC.

The SOC is then defined by

$$SOC = 100 \cdot \frac{C_{cell} - C_{discharged}}{C_{cell}} \%. \tag{2.13}$$

All steps are conducted at nominal temperature given by the cell producer. For the experimental cells mainly used in this thesis, the SOC is always related to the real capacity of the corresponding cell.

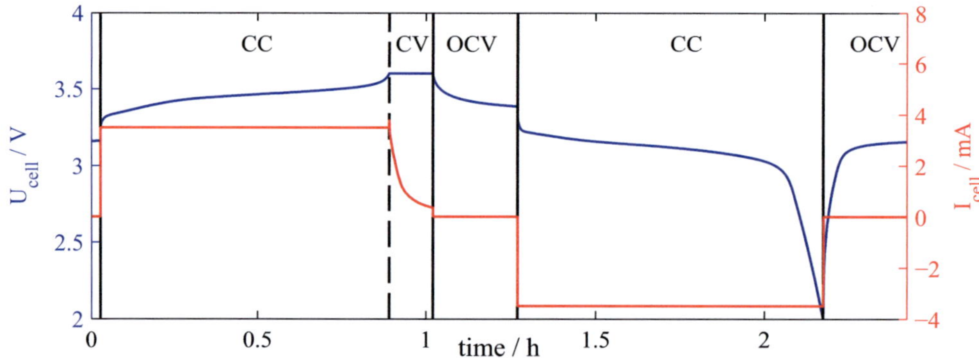

Figure 2.10: Standard charge (first CCCV phase) and discharge (second CC phase) procedure for an experimental $LiFePO_4$/graphite cell.

2.1.5.5 Full Cell Capacity and Electrode Balancing

The capacity of lithium-ion full cells is given by the capacity of its electrodes and the active lithium which is available in the cell for cycling. The lithium is stored in the cathode before the cell fabrication due to safety reasons. Therefore, the amount of active lithium introduced into the cell is limited by the cathode capacity. Furthermore, a part of the lithium is consumed during the formation of the SEI, leading to a decreasing capacity. Consequently, only a part of the cathode capacity is used during cycling. This has also an impact on the cell voltage, as the cathode can no longer be completely lithiated after the SEI formation, leading to a higher cathode voltage.
Another factor which influences the cell capacity and the cell voltage is the relation between cathode and anode capacity, known as *balancing* or *matching*. Usually, the anode capacity is significantly higher than the cathode capacity. This is done for safety and degradation reasons. The insertion of too much lithium into graphite-anodes (more than LiC_6) causes the deposition of lithium metal on the graphite surface which is

highly reactive, leading to the same risk as with the lithium metal anodes previously introduced.

All these factors affect the cell capacity and also the characteristic SOC dependency of the OCV. Figure 2.11 demonstrates the correlation between electrode balancing, active lithium available and the cells' characteristic OCV curve. The gray bars indicate the area of electrode capacity not used in the simulated cell configuration. In the present case, a large part of the graphite-anode is not used but only a small part of the cathode capacity is not used. Depending on the capacity of both electrodes, the voltage profiles are shifted and scaled versus each other and another cell voltage emerges [56]. The OCV curve varies during cell degradation because the capacity of the active materials or the amount of active lithium may change.

Knowing the electrode balancing is important for the separate electrode analysis in order to know the single electrode potentials which correspond to specific full cell SOCs. This enables the analysis of single electrodes in half-cells at a comparable SOC.

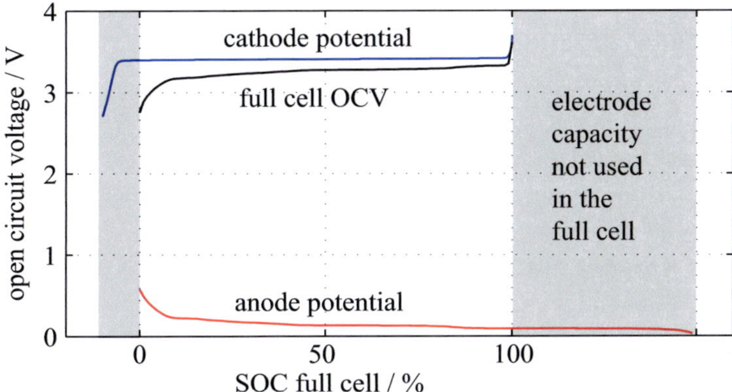

Figure 2.11: Full cell voltage and electrode potentials for a LiFePO$_4$/graphite cell.

2.1.5.6 Coulombic Efficiency

The coulombic efficiency η_{Ah} is defined for the charge/discharge cycle of a lithium-ion cell by

$$\eta_{Ah} = \frac{Q_{discharge}}{Q_{charge}}. \tag{2.14}$$

This is a measure for the lithium consumption during the cycling of a cell. It is lower during SEI-formation in the first circles, whereas it increases up to 99.9 % afterwards for state-of-the-art lithium-ion cells [57]. This is very important as a low coulombic efficiency causes an irreversible capacity loss.

2.2 Electrochemical Impedance Spectroscopy

Electrochemical Impedance Spectroscopy (EIS) is a well-established tool for the investigation of electrochemical systems [53,58]. The latter has to fulfill the conditions of causality, linearity and time invariance in order to allow for a meaningful analysis. The goal is an identification of electrochemical mechanisms, denominated as loss processes, responsible for performance limiting overpotentials. This section will address the general measurement principle and several possibilities to analyze the impedance data obtained.

2.2.1 Measurement Principle

Electrochemical impedance spectroscopy is based on the fact that electrochemical loss processes occur in a wide range of frequencies, each process having its own characteristic time constant [51,59,60]. The determination of a frequency dependent complex impedance spectrum enables the identification and separation of these loss processes. For that purpose, a sinusoidal excitation signal (voltage $u\left(t\right) = u_{amp}\left(\omega\right) \cdot sin\left(\omega t\right)$ or current $i\left(t\right) = i_{amp}\left(\omega\right) \cdot sin\left(\omega t\right)$ with $\omega = 2\pi f$) is applied and the resulting system response $\left(i\left(t\right) = i_{amp}\left(\omega\right) \cdot sin\left(\omega t + \varphi\left(\omega\right)\right)\right.$ or $\left.u\left(t\right) = u_{amp}\left(\omega\right) \cdot sin\left(\omega t + \varphi\left(\omega\right)\right)\right)$ is measured (see Figure 2.12a). Dividing the voltage by the current using Equation 2.15

$$\underline{Z}\left(\omega\right) = \frac{u_{amp}\left(\omega\right)}{i_{amp}\left(\omega\right)} \cdot e^{j\varphi(\omega)} = \left|\underline{Z}\left(\omega\right)\right| \cdot e^{j\varphi(\omega)} = Z'\left(\omega\right) + j \cdot Z''\left(\omega\right) \tag{2.15}$$

yields the complex impedance $\underline{Z}\left(\omega\right)$. It is defined by its absolute value $\left|\underline{Z}\left(\omega\right)\right|$ and the phase shift $\varphi\left(\omega\right)$ between the two sinusoidal signals, illustrated in Figure 2.12. Another representation is given by real ($Z'\left(\omega\right)$) and imaginary ($Z''\left(\omega\right)$) part of the complex impedance. Performing this measurement for several frequencies leads to the complex impedance spectrum shown by example of a lithium-ion cell in Figure 2.12b. As most electrochemical systems show a capacitive behavior, the imaginary axis is flipped vertically, here and in all following figures, in order to plot the spectrum in the upper half-plane. For a meaningful determination of complex impedance spectra it is necessary to chose several parameters carefully. First, the amplitude of the stimulation signal has to be small enough to stay in a linear region of the operating point. This is very important as the lithium-ion cell is by nature a non-linear system. Amplitudes between $5\,\mathrm{mV}$ and $10\,\mathrm{mV}$ have proven to be a good value for potentiostatic measurements where the applied signal is a sinusoidal voltage. For galvanostatic measurements (a sinusoidal current is applied), obtaining a voltage response around $10\,\mathrm{mV}$ is desirable as well. Therefore, the current amplitude has to be chosen according to the absolute value $\left|\underline{Z}\left(\omega\right)\right|$ of the impedance. Further, some measurement equipment provides a pseudo-potentiostatic method where a target amplitude of the voltage is set and the impedance measurement is conducted in a galvanostatic mode with a current amplitude automatically chosen.

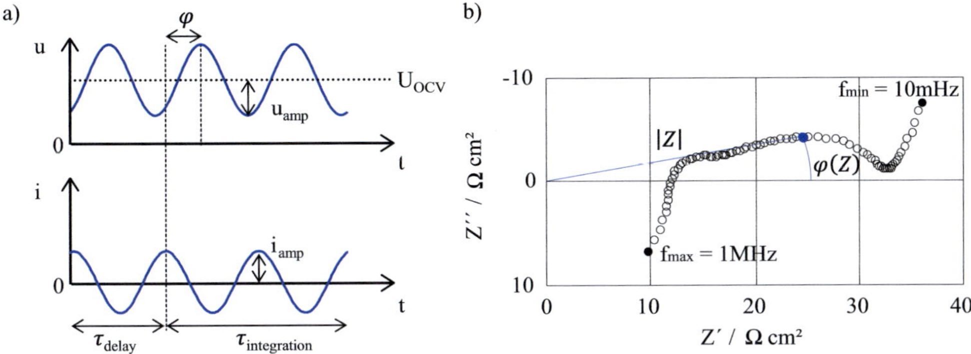

Figure 2.12: a) Sinusoidal voltage and current signal during an impedance measurement. b) Nyquist plot of a typical impedance spectrum for an experimental lithium-ion cell introduced in Section 4.1.

In theory, the response of a linear system to a stimulation signal with one specific frequency contains only this specific frequency. The response of a non-linear system would contain contributions of other harmonics as well [61]. Transforming the signal by Fourier transform allows for the detection of these harmonics and the underlying non-linearity. A common way for finding out if an amplitude is small enough, is to check whether the amplitude changes the impedance spectrum. If so, the chosen amplitude is to high. If not, it is small enough to be considered a linear system.

Another important parameter is the integration time for the excitation signal. For each frequency point, the first part of the stimulation is used as a delay time which is not evaluated in order to obtain a steady system as illustrated in Figure 2.12a. Afterwards, the signal is evaluated for a certain time or a certain number of oscillation periods from which the complex impedance is calculated. The choice of integration time is a compromise between good quality of the measurement data and a desired short measurement time. The Kramers Kronig residuals are calculated in order to mathematically assess the measurement data quality (Section 2.2.2).

In general, it is possible to add an offset current during a galvanostatic impedance measurement. However, for lithium-ion batteries this is not common as the offset current changes the SOC and violates the condition of a time-invariant system. Another problem area is the measurement of impedance spectra to very low frequencies. First, the charge inserted increases with decreasing frequency. This increases the ΔSOC during a measurement and therefore changes the lithium-ion cell during the measurement. Moreover, the measurement time increases disproportionally and increases the aging of the investigated cell during the measurement. Both cases cause a time-variant system. Therefore, complementing impedance measurements by time domain measurements in order to asses lower frequencies is recommendable (see Section 2.3.2 or Ref. [62]).

2.2.2 Kramers Kronig Residuals

A meaningful evaluation of impedance spectra requires high measurement data quality and meeting the conditions mentioned above. The Kramers Kronig (KK) residuals have proven to be a suitable method to check the measurement data quality and the condition of time invariance for impedance measurements. They are based on the relation between real and imaginary parts for a linear time invariant system, given by the following equations [58, 63, 64]:

$$Z'(\omega) = Z'(\infty) + \frac{2}{\pi} \cdot \int_0^\infty \frac{xZ''(x) - \omega Z''(\omega)}{x^2 - \omega^2} dx, \qquad (2.16)$$

and

$$Z''(\omega) = -\frac{2\omega}{\pi} \int_0^\infty \frac{Z'(x) - Z'(\omega)}{x^2 - \omega^2} dx. \qquad (2.17)$$

The problem with these equations is the integration over the frequency range from 0 to ∞ as impedance measurements never cover this frequency range. An approach to overcome these limitations was initially proposed by Agarwal et al. [65]. It is based on a series connection of RC-elements (the measurement model), known to be KK-transformable. The latter is used to fit the real or the imaginary part of the measured impedance data by a Complex Non-linear Least Squares fit (CNLS-fit) and to predict the other component. The relative deviation of the predicted from the measured values is called the KK residual. It can be used to measure whether or not the impedance data is KK-transformable. The approach was improved by Boukamp et al. [64]. Therein, the time constants of the RC-elements are fixed and only the polarization is a free fit parameter. A complex linear least squares fit results and allows for a more stable fit of the impedance data. It can be further complemented by a series connected inductor and/or capacitance in order to describe pure capacitive or pure inductive behavior [64] as it occurs in lithium-ion cells.

2.2.3 Distribution of Relaxation Times

The concept of the Distribution of Relaxation Times (DRT) was introduced for dielectrics and capacitors in 1907 by Schweidler [66] and further developed by Wagner in 1913 [67]. A first successful application of the DRT to energy converting systems was introduced by Schichlein et al. in Refs. [68, 69] for Solid Oxide Fuel Cells (SOFC). They demonstrated that an identification of loss mechanisms with higher frequency resolution compared to standard impedance analysis could be obtained. Further, the DRT allowed for a better visualization of individual parameter dependencies. In Refs. [70, 71] the DRT was successfully applied for the design of a physically motivated impedance model for the SOFC.

2.2.3.1 Concept

In the following section, the concept of the DRT and its implementation will be introduced. The DRT-method is based on the fact that every impedance function that obeys the KK relations can be represented by an infinite number of infinitesimal small differential RC-elements [53]. This is common in system theory and also holds true for electrochemical systems like lithium-ion batteries. The relation between the distribution function $\gamma(\tau)$ and the impedance $\underline{Z}(\omega)$ is given by the following equation:

$$\underline{Z}(\omega) = R_0 + R_{pol} \cdot \int_0^\infty \frac{\gamma(\tau)}{1 + j\omega\tau} d\tau \tag{2.18}$$

with

$$\int_0^\infty \gamma(\tau) \, d\tau = 1. \tag{2.19}$$

The term $\frac{\gamma(\tau)}{1+j\omega\tau} d\tau$ specifies the fraction of the overall polarization with relaxation times between τ and $\tau + d\tau$. R_0 represents the ohmic resistance and R_{pol} the overall polarization. The angular frequency is defined as $\omega = 2\pi f$.

In practice, a finite number of N RC-elements is used for a numerical DRT approximation given by the discrete distribution function g_n. The corresponding equation is expressed by

$$\underline{Z}(\omega) = R_0 + R_{pol} \cdot \sum_{n=1}^N \frac{g_n}{1 + j\omega\tau_n} = R_0 + \sum_{n=1}^N \frac{R_n}{1 + j\omega\tau_n} \tag{2.20}$$

where g_n is the contribution of the nth RC-element with the specific time constant τ_n to the overall polarization. It has proven to be advantageous for the visualization of the DRT to chose the time constants τ_n logarithmically distributed and to plot the DRT versus the logarithmic frequency [69]. Furthermore, the product of polarization and density $R_{pol} \cdot g$ is usually depicted instead the density g. Yet the DRT is denominated as g in all figures of this thesis, following the literature [70–72]. The unit of the DRT is $[\Omega s]$ and for the area specific DRT $[\Omega scm^2]$. Due to the connection of the real part and the imaginary part by the KK relation it is possible to calculate the DRT only from the real or the imaginary part of the impedance spectrum.

2.2.3.2 Calculation from Measured Impedance Data

The solution of Equation 2.20 represents an inverse ill-posed problem. It can be approximated numerically with the help of regularization. In this thesis, a Tikhonov regularization [73] is applied in accordance to literature [70–72, 74] in order to obtain a stable solution. The curvature of the DRT function $g(f)$ is applied as an input parameter for the regularization. It is multiplied with a weight factor λ. A higher value for λ increases the weight of the curvature, leading to a smoothing out of the DRT function. The impact of the regularization parameter λ on the DRT calculation is clarified by Figure 2.13. Herein, the DRT for an RC-element is calculated using different

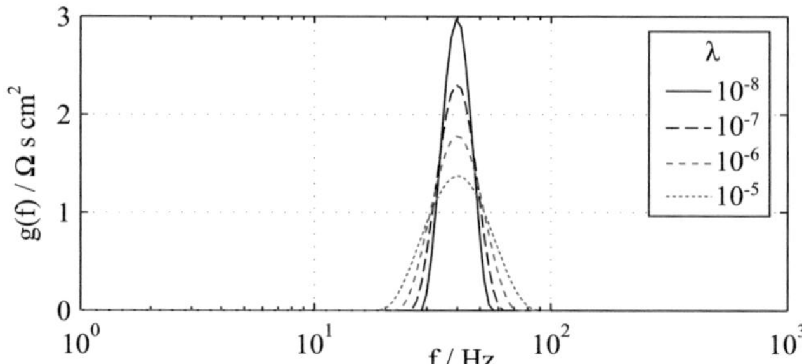

Figure 2.13: DRT of an ideal RC-element for varying regularization parameter λ (R_{RC} =1 Ωcm^2, τ_{RC} =4 ms).

regularization parameters λ. In theory, the DRT of an RC-element is represented by a dirac peak at its characteristic frequency $f_{char} = \frac{1}{2\pi\tau_n}$. In contrast, the numerical DRT-calculation yields a broader peak with a maximum at the characteristic time constant. The area enclosed by the peak corresponds to the polarization of the underlying process. The peak shape broadens further for increasing λ as the influence of the DRT curvature is thereby increased.

In case of real systems, a strong regularization can cause a merge of peaks with closed characteristic time constants and therefore prevent a separation of the underlying processes. In contrast, additional peaks caused by measurement noise may occur for small λ. These false peaks are called artifacts and may interfere with the identification of real processes [74]. The adequate selection of λ and a high measurement data quality are obviously crucial factors for DRT analysis.

2.2.4 Complex Non-Linear Least Squares Fit

After having identified the dominant electrochemical loss mechanisms by DRT analysis, it is possible to develop suitable Equivalent Circuit Models (ECM). The latter should be able to describe the impedance data properly and, in the same time, have a physical meaning. These models comprise of several serial impedance elements or more complex model structures, introduced in Section 2.4. The impedance model thus developed can subsequently be applied to evaluate the measured impedance spectra by a CNLS-fit. Here, the relative deviation between the non-linear model function and the measured impedance spectrum is minimized. The quality criterion used in this thesis is calculated from the real and the imaginary part by the following equation:

$$S = \sum_{k=1}^{N} \left[w_{k,real} \frac{\left(Z'\left(\omega_k\right)_{mod} - Z'\left(\omega_k\right)_{meas}\right)^2}{\left|Z\left(\omega_k\right)\right|_{meas}} + w_{k,im} \frac{\left(Z''\left(\omega_k\right)_{mod} - Z''\left(\omega_k\right)_{meas}\right)^2}{\left|Z\left(\omega_k\right)\right|_{meas}} \right].$$

$$(2.21)$$

$Z'(\omega_k)_{mod}$ and $Z''(\omega_k)_{mod}$ denote the real and the imaginary part of the model and $Z'(\omega_k)_{meas}$ and $Z''(\omega_k)_{meas}$ those of the measured impedance at the angular frequency ω_k. The weight factors $w_{k,real}$ and $w_{k,im}$ are set to one if no systematic difference in measurement data quality is identified for the frequency range measured.

2.2.4.1 Relative Fit Residuals

The fit quality can be visualized by the relative residuals of measured and simulated impedance spectra defined by the following equations for the real part

$$res_{real}(\omega) = \frac{Z'(\omega)_{mod} - Z'(\omega)_{meas}}{|Z(\omega)_{meas}|} \tag{2.22}$$

and the imaginary part

$$res_{im}(\omega) = \frac{Z''(\omega)_{mod} - Z''(\omega)_{meas}}{|Z(\omega)_{meas}|}. \tag{2.23}$$

These residuals represent a mathematical measure for the fit quality. Two kinds of fit residuals may occur during evaluation assuming small, randomly distributed KK residuals:

1. In the first case, small randomly distributed model residuals are obtained. This indicates a good fit quality where the remaining residuals are caused by the measurement noise already detected in the KK residuals. The model is therefore able to describe the impedance data correctly. This mathematical property is not yet a measure for the physical meaning of the impedance model. For an evaluation of the physical suitability of an impedance model, it is necessary to evaluate impedance models for various operating parameters. The model must be able to describe the impedance for all parameters, providing physically meaningful parameter dependencies for each loss process.

2. In the second case, systematically distributed residuals occur. This demonstrates a bad fit quality due to incorrect starting parameters or a fundamentally incorrect impedance model. The risk for both can be reduced by using the DRT method introduced above. The required number and type of impedance model elements is then identified by DRT analysis and the choice of starting parameters is also supported. The characteristic frequency of each loss process is given directly by the maximum of the corresponding peak, whereas its polarization can be estimated from the peak area.

The relative residuals and the obtained parameter dependencies yield a good base for the assessment of an impedance models' suitability. However, the comparison of a high number of impedance measurements or a high number of impedance models by relative residuals is confusing. Therefore, the residuals can be used to calculate another measure for fit quality.

2.2.4.2 Root Mean Squares Error

The Root Mean Squares Error (RMSE) can be calculated directly from relative fit residuals using Equation 2.24.

$$RMSE = \sum_{k=1}^{N} \left[\frac{\sqrt{\left(Z'\left(\omega\right)_{mod} - Z'\left(\omega\right)_{meas}\right)^2 + \left(Z''\left(\omega\right)_{mod} - Z''\left(\omega\right)_{meas}\right)^2}}{\left|Z\left(\omega\right)\right|_{meas}} \right] / N \tag{2.24}$$

The RMSE does not show whether systematically distributed or randomly distributed fit residuals occur and therefore it cannot replace the analysis of the residuals. However, for the evaluation of a high number of fit results it can be helpful.

2.3 Time Domain Measurements

Time domain measurements (TDM) are based on the application of current/voltage profiles different from sinusoidal excitation signals. In the simplest case, current or voltage pulses or steps are applied. There are four common methods to evaluate TDMs. In the simplest approach, not discussed in detail in this thesis, the voltage drop during pulse discharge or charge is used in order to calculate a DC resistance. Details about this method can be found in Refs. [74–76]. This method yields only the overall polarization during charge or discharge without distinguishing different loss mechanisms or the effect of temperature rise.

The other methods which are more elaborate will be introduced in the following two sections.

2.3.1 GITT and PITT

Galvanostatic Intermittent Titration Technique (GITT) and Potentiostatic Intermittent Titration Technique (PITT) are two time domain measurement methods presuming the solid state diffusion to be the major contribution at low frequencies. Both methods enable the determination of diffusion coefficients.

GITT is based on the application of current pulses to the electrochemical system investigated. Figure 2.14 shows a schematic voltage response on a current pulse and the parameters needed for the evaluation. Assuming a diffusion process describable by Ficks' law leads to the following equation for the determination of the diffusion coefficient [57, 77]:

$$D = \frac{4}{\pi \tau_{pulse}} \left(\frac{m_b V_{Molar}}{M_b S} \right)^2 \left(\frac{\Delta E_s}{\Delta E_t} \right)^2 \tag{2.25}$$

Here, m_b is the mass, V_{Molar} the molar volume, M_b the molar mass and S the active surface are of the active material. ΔE_s, ΔE_t and τ_{pulse} are determined from the voltage response. This approach is only valid if the pulse duration τ_{pulse} is kept short enough: $\tau_{pulse} << \frac{L^2}{D}$.

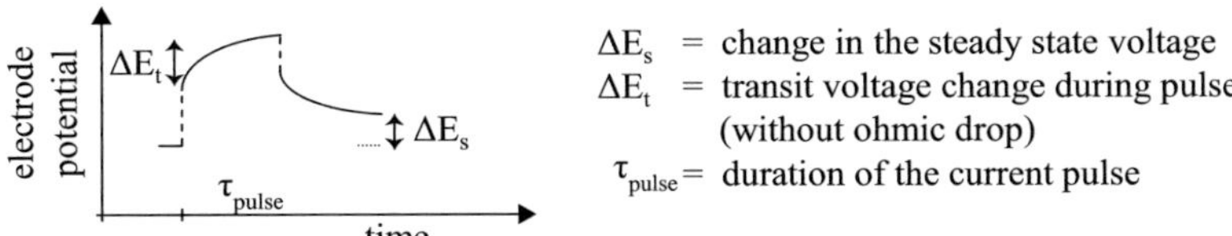

Figure 2.14: Schematic voltage response on a current pulse applied for the GITT method.

An alternative method using the same presumptions as above is provided by the PITT method [57, 78]. Here, a small amplitude voltage step is applied to the electrode and the resulting current is recorded. A typical current response is illustrated in Figure 2.15.

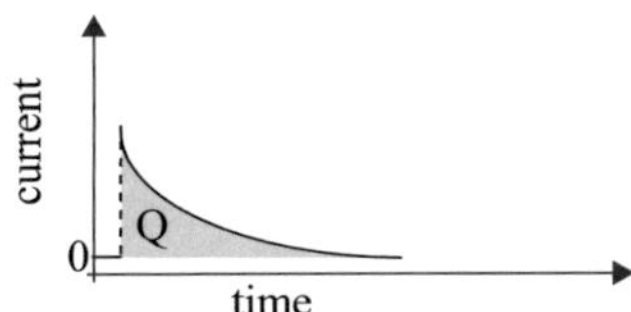

Figure 2.15: Current response on a voltage step applied for the PITT method.

There are two approaches based on the diffusion length and the time required to let the current abate during the voltage step. Both equations are provided in the following:

$$I(t) = \frac{QD^{1/2}}{L\pi^{1/2}} \left(\frac{1}{t^{1/2}} \right) \quad \text{for } t << L^2/D \tag{2.26}$$

and

$$I(t) = \frac{2QD}{L^2} exp \left(\frac{-\pi^2 Dt}{4L^2} \right) \quad \text{for } t >> L^2/D \tag{2.27}$$

Both approaches need a different evaluation of the current response. For the first approach, the current $I(t)$ is plotted against $\frac{1}{t^{1/2}}$. For the second approach, $log(I(t))$ is plotted against the time. A calculation of the diffusion coefficient is then possible by the determination of the slope of each current function and by the use of Equations 2.26 and 2.27 respectively.

GITT and PITT are derived from a one-dimensional diffusion geometry. For real electrodes, this assumption is usually not true as they have a porous structures and they contain particles with a particle size distribution. The solid state diffusion into these different particles shows different time constants and the porous structure causes potential distributions in the electrode and therefore a distributed lithium insertion. Due to these factors it is only possible to determine effective diffusion coefficients in real electrodes; however, the latter are not equivalent to real solid state diffusion coefficients. A detailed investigation of GITT for the application to porous electrodes is presented in Ref. [79].

2.3.2 Pulse-Fitting

Pulse-fitting is a method used to calculate impedance spectra from a current or voltage pulse or step and the corresponding response of the electrochemical system under investigation. It was first introduced in Ref. [74] and its fundamentals will be explained accordingly in the following section. The main advantage of this approach is its capability to analyze lower frequencies compared to standard electrochemical impedance spectroscopy in the same measurement time. Furthermore, it enables a non-parametric evaluation of very low frequencies with less a-priori assumptions compared to GITT/PITT. A more general approach to calculate the impedance response from a TDM would be a Fourier transformation of the measured data. With this method it is possible to calculate the impedance spectrum from current/voltage pulses [62] and also from the response on arbitrary excitation signals.

2.3.2.1 Measurement Principle

The pulse-fitting approach applied in this thesis is based on the application of discharge current pulses. Each measurement consists of four steps:

1. SOC setting
2. voltage relaxation phase I
3. current pulse
4. voltage relaxation phase II

Both voltage relaxation phases should be of similar length in order to enable an abatement of the overpotentials caused by setting the SOC. The analysis of the voltage response is based on the equivalent circuit model presented in Figure 2.16, which can be assumed for any capacitive system. It consists of an ohmic resistance and a serial connection of several RC-elements in order to describe the overpotentials and of a charged capacity $C_{\Delta int}$ to represent the cell voltage U_{OCV} and the capacitive behavior. Accordingly, the voltage response due to a current pulse can split up like the different components of the model (Figure 2.17).

First, the overpotential due to the ohmic resistance (ohmic drop) occurs instantaneously after the current pulse has started. Similarly, it disappears when the current has been interrupted. Second, the serial connection of RC-elements causes a dynamically

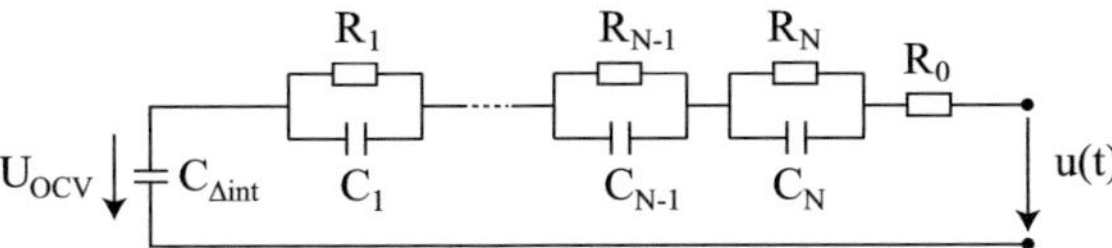

Figure 2.16: Equivalent circuit model used for the pulse-fitting approach consisting of ohmic resistance, RC-elements and intercalation capacity.

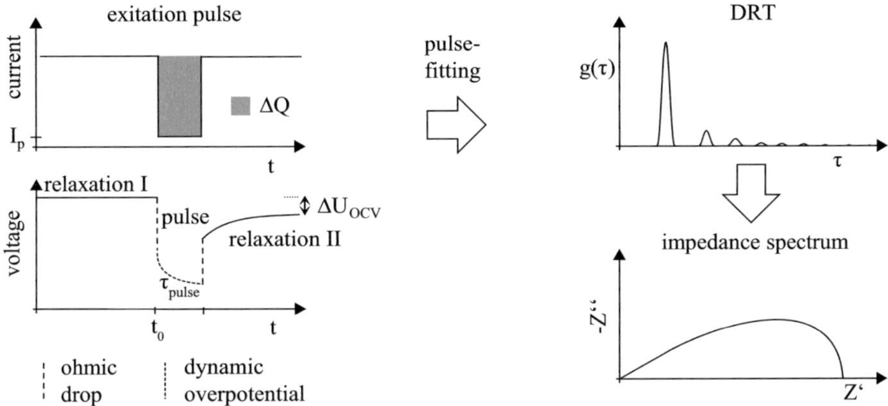

Figure 2.17: Steps to calculate the impedance spectrum from TDM by pulse-fitting.

increasing overpotential according to their time constants τ_n and resistance values R_n. Again, this overpotential reduces dynamically after the current pulse has been interrupted. Finally, the cell voltage converges at the end of the voltage relaxation to a stationary value which corresponds to the OCV of the specific SOC. This voltage is usually different from the voltage before the pulse, depending on the slope of the OCV curve at the SOC investigated. This voltage difference characterizes the intercalation capacity $C_{\Delta int}$, which can be calculated from the OCV difference ΔU_{OCV} and the discharge capacity ΔQ by the following equation:

$$C_{\Delta int} = \frac{\Delta Q}{\Delta U} \tag{2.28}$$

By this definition, the intercalation capacity is the reciprocal value of the slope of the OCV curve. Consequently, it depends on the shape of the OCV curve and is SOC dependent. It can be measured by pulse charge or discharge experiments, by potential steps [57] or by calculating the reciprocal of the derivation from slow charge/discharge curves [80]. It can also be determined from impedance measurements as it corresponds to the capacitive behavior of the impedance spectra.

The goal of the pulse-fitting approach is the determination of the parameters of the equivalent circuit model introduced above which are suitable to describe the system optimally. The steps needed for that are comparable to the calculation of the DRT from impedance spectra, suggesting the application of similar methods. The first step is to decide a number of RC-elements used. Similar to the DRT, it is obvious that a high number of RC-elements increases the accuracy of the model but also expands the computing time. Next, the time constants of these RC-elements are fixed and logarithmically distributed over the frequency range investigated. The latter is restricted theoretically for small time constants by the sample rate due to the Nyquist-Shannon sampling theorem. In Ref. [74] it is recommended to chose the minimum time constant τ_{min} according to:

$$\tau_{min} = \frac{100 \cdot 2}{2\pi f_s} \tag{2.29}$$

Here f_s denominates the sample rate of the measurement. Furthermore, the maximum time constant τ_{max} is restricted according to Equation 2.30 by the relaxation time τ_{relax} after the current pulse.

$$\tau_{max} \leq \tau_{relax} \tag{2.30}$$

The next step after choosing the time constants is the estimation of R_n for all RC-elements. It starts with the set-up of the theoretical equation for the voltage response of an RC-element on a current pulse. Equation 2.31 describes the voltage relaxation after the pulse.

$$u_n(t) = R_n I_p \left[\left(1 - e^{-\frac{t-t_0}{\tau_n}}\right) - \left(1 - e^{-\frac{t-\left(t_0+\tau_{pulse}\right)}{\tau_n}}\right) \right] \overset{t=t_j}{=} R_n \cdot a_{nj} \tag{2.31}$$

Here, t_0 is the starting time of the current pulse, τ_{pulse} the pulse length, τ_n the time constant of the RC-element and I_p the current amplitude. The only unknown parameter for the determination of $u_n(t_j)$ is the resistance R_n if a discrete time vector t_j is inserted. The equation can be reformulated into a linear equation containing a simple multiplication of parameter $a_n(t_j) = a_{nj}$ and resistance R_n. This linear formulation allows for the set-up of the system of linear equations presented in Equation 2.32.

$$\mathbf{AR} = \mathbf{u} - U_{OCV} \tag{2.32}$$

$\mathbf{R}$ is an N-dimensional vector containing R_n for all RC-elements, $\mathbf{A}$ is a $M\mathrm{x}N$ matrix of the parameters $a_{n,j}$ with $n = 1 \ldots N$ (N =number of RC-elements) and $j = 1 \ldots M$ (M =number of measurement points). $\mathbf{u}$ is an M-dimensional vector of the cell voltages for discrete time values t_j and U_{OCV} the open circuit voltage of the cell.

Equation 2.32 is then rearranged in order to include U_{OCV} into the parameter estimation. The parameters R_n and U_{OCV} are finally approximated by a least squares estimation. For that, a Tikhonov regularization [73] is applied in order to stabilize the solution. The curvature multiplied with a weight factor λ (regularization parameter) is then used as additional quality criterion for the estimation. This leads to a vector of resistances

R_n corresponding to the DRT, but directly calculated from time domain data. For the following chapters, the DRT from TDM will be plotted against the angular frequency $\omega = \frac{1}{2\pi\tau}$ instead of the time constants. Further details about the entire procedure can be found in Ref. [74].

2.3.2.2 Compensation for Self-Discharge

During the relaxation phase of the TDM previously introduced, the SOC is not changed actively and should remain constant as no current is applied. However, a change of the OCV can be observed for real lithium-ion cells during voltage relaxation. In the case of a discharge pulse, it can be identified by a decreasing cell voltage at the end of the relaxation phase. Such behavior is known as self-discharge [74, 81–83] and occurs to a different extend depending on cell type, temperature and SOC. It is observed for most lithium-ion cells and can be caused by several mechanisms such as electronic electrolyte conductivity, SEI formation on the anode side [81, 83] or other parasitic reactions in the electrolyte and its interfaces to the electrodes [82].
A significant self-discharge impedes the evaluation of TDMs with the previously introduced procedure as a decreasing potential cannot be described by the equivalent circuit model applied. Therefore, it is necessary to detect and compensate for it in order to obtain a meaningful evaluation. The detection of self-discharge can be done by a very simple approach for discharge pulses which is the main reason why no charge pulses are applied. After a discharge pulse, the cell voltage shows an increasing voltage until the self-discharge starts to exceed the voltage relaxation. From this point of maximum voltage, a decreasing voltage is observed. This can be used to its advantage by simply searching the maximum voltage after the current pulse and comparing it to the last cell voltage measured. A relevant self-discharge is detected if the voltage gradient between maximum voltage and last measurement point is below a certain value, calculated from the resolution ϵ of the measurement equipment according to Equation 2.33 (Ref. [74]).

$$\Delta U_{min} = -\epsilon \cdot \frac{1}{10 \cdot 3600s} \tag{2.33}$$

In this case, a compensation is necessary prior to the application of the pulse-fitting. The region of dominating self-discharge is identified by dividing the voltage relaxation phase into several segments and comparing their voltage gradients. Starting from the latest measurement segment, the gradients are compared sequentially between two neighboring segments for the entire relaxation phase. A difference in the gradient of more than ten percent marks the end of the self-discharge region. This method assumes a linear self-discharge which does not change its intensity.
A linear self-discharge can be described by a simple resistance R_{self} connected in parallel to the lithium-ion cell. It can be determined by a fit of the resulting voltage drop to the region of dominating self-discharge previously identified. Afterwards, the corresponding voltage drop is subtracted from the measured voltage in order to prevent

an effect on the pulse-fitting evaluation. The self-discharge model and compensation are explained in great detail in Ref. [74].

2.3.2.3 Combining EIS and TDM

With the time domain approach presented in the previous sections it is not possible to determine the ohmic resistance and the high frequency impedance spectrum for characteristic time constants below τ_{min} (see Equation 2.29). A combination of TDMs with standard electrochemical impedance spectroscopy is therefore recommended in order to cover a larger frequency range. A merge of both impedance spectra has to be conducted in order to obtain a correct impedance spectrum for all frequencies. The merge is accomplished by adjusting the real part of the time domain impedance spectrum according to the gap between both impedance spectra at the last frequency point of the standard impedance spectrum:

$$R_{offset} = Z'_{EIS}(f_{min}) - Z'_{TD}(f_{min}) \tag{2.34}$$

The impedance spectrum from TDM is then shifted along the real axis by R_{offset} in order to merge both impedance spectra.

2.4 Equivalent Circuit Models

Equivalent circuit models (ECM) are a well-established tool for the analysis of impedance spectra of electrochemical systems [53]. They allow for the quantification of individual contributions of different loss mechanisms and the identification of their parameter dependencies. This section gives an overview of the equivalent circuit elements used in this thesis and the model structures to connect these elements.

2.4.1 Basic Elements

ECMs consist of several impedance elements which are serially connected or connected in more complex model structures. In this section, several of these simple impedance elements are introduced. From electrical networks the basic elements such as resistor, inductor and capacitor are well-known. These elements can be applied for the modelling of electrochemical systems as well. Their complex frequency dependent impedance needed for impedance modelling is expressed by

$$Z_R = R \qquad Z_C(\omega) = \frac{1}{j\omega C} \qquad Z_L(\omega) = j\omega L \tag{2.35}$$

with angular frequency $\omega = 2\pi f$ and imaginary unit j. The resistor is frequency independent whereas capacitor and inductor show a frequency dependent imaginary part.

2.4.1.1 RC-element

The parallel connection of resistance and capacitance – called RC-element – is one of the most common elements in electrochemical impedance modelling. As already explained in Section 2.2.3, it is possible to describe every causal, linear and time invariant system by a serial connection of RC-elements. However, in electrochemistry the use of RC-elements can be more physically motivated compared to this generalized approach. Its impedance is given by the following function:

$$Z_{RC}(\omega) = \frac{R}{1 + j\omega RC} = \frac{R}{1 + j\omega\tau} \tag{2.36}$$

where R represents the polarization and C the capacity of the RC-elements. The reformulation using the time constant $\tau = RC$ gives more graphic description as it specifies the characteristic time constant τ of the RC-element directly. The impedance spectrum described by RC-elements represents an ideal semicircle in the Nyquist plot. In electrochemistry, RC-elements are usually applied for the description of interface loss processes, for instance a charge transfer process between electrode and electrolyte. The resistance represents the limited exchange rate, whereas the capacitor describes the double layer capacity. The use of a simple RC-element in order to describe the charge transfer is only valid if the diffusion of species to the reaction site can be neglected and if the geometry of the reaction site is an ideal and plain surface structure.

2.4.1.2 RQ-element

In real electrochemical systems it is very rare for clear RC behavior to be observed. It is more common that depressed semicircles occur which cannot be described by one RC-element. In that case, RQ-elements are more suitable in which the capacitance in the RC-element is replaced by a Constant Phase Element (CPE) [84]:

$$Z_{CPE}(\omega) = \frac{1}{A_0 (j\omega)^n} \tag{2.37}$$

with an exponent $0 < n < 1$. For $n = 1$ the CPE is equal to a capacitor, for $n = 0.5$ it is equal to an infinite solid state diffusion process and for $n = 0$ it corresponds to a resistor [53]. The CPE is an empirical element without a direct physical meaning. Possible explanations for constant phase behavior are given in Refs. [85–87]. Its impedance shows capacitive behavior and an increasing real part for decreasing frequencies (Figure 2.18a).

Connected in parallel with a resistor, the CPE leads to the RQ-element also known as the ZARC-element [88] and in analogy to the Cole-Cole element [89].

$$Z_{RQ}(\omega) = \frac{R}{1 + RA_0 (j\omega)^n} = \frac{R}{1 + (j\omega\tau)^n}. \tag{2.38}$$

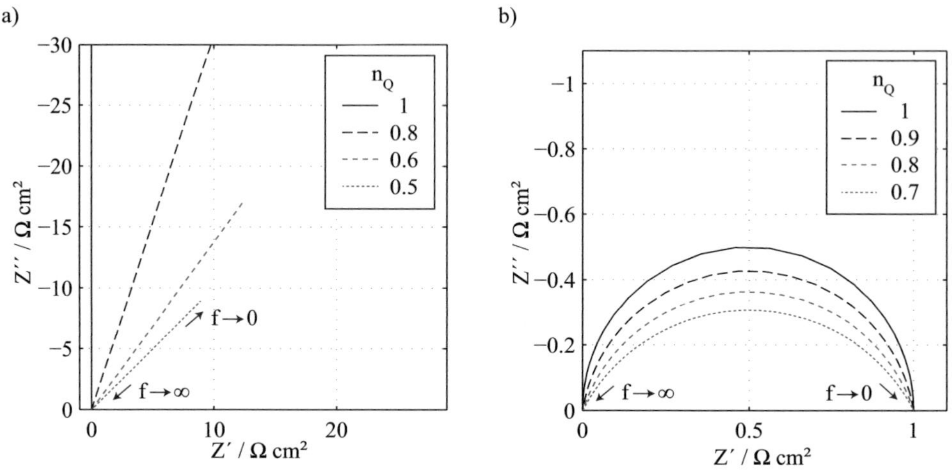

Figure 2.18: a) Nyquist plot of CPE-element for varying n_Q ($A_0 = 1\Omega cm^2$). b) Nyquist plot of RQ-element for varying n_Q ($R_{RQ} = 1\,\Omega cm^2$, $\tau_{RQ} = 4\,ms$).

For $n = 1$ the impedance is equal to that of an RC-element. For lower exponent values, it represents depressed semicircles (Figure 2.18b). This originates from a distribution of time constants around a peak time constant which corresponds to the characteristic time constant of the RQ-element (Figure 2.19). Due to the numerical calculation of the DRT a wide peak results even for $n = 1$ instead of a dirac [71]. The peak width increases with decreasing exponent n. A direct relation between exponent n and the physical origin thereof is not given [90]. However, a distribution of time constants for real electrochemical processes can be related to distributed parameters due to microstructure effects. For instance varying particle size in a composite electrode, surface heterogeneity or different reaction sites due to inhomogeneities in the lattice structure can be given [58].

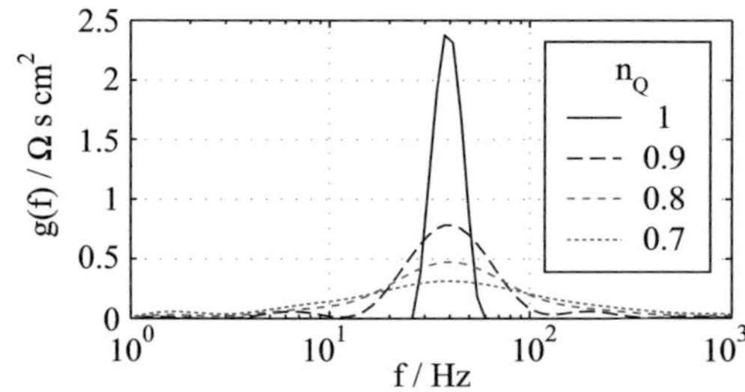

Figure 2.19: DRT of RQ-element for varying n_Q ($R_{RQ} = 1\,\Omega cm^2$, $\tau_{RQ} = 4\,ms$, $\lambda_{DRT} = 10^{-6}$).

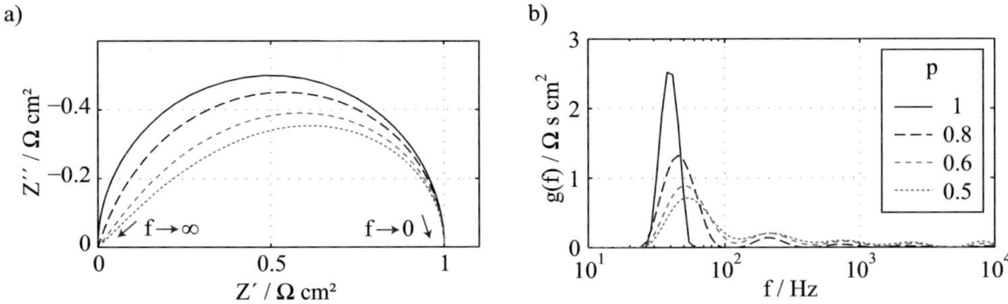

Figure 2.20: a) Nyquist plot and b) DRT of HN-element for varying p_{HN} ($R_{HN} = 1\,\Omega\text{cm}^2$, $\tau_{HN} = 4\,\text{ms}$, $n_{HN} = 1$, $\lambda_{DRT} = 10^{-6}$).

2.4.1.3 HN-element

Havriliak and Negami introduced empirical equations yielding an equivalent circuit element (HN-element) in order to describe relaxation processes in polymers [91, 92]. In Ref. [93] these HN-equations were applied to describe dielectric relaxation in lithium-ion conducting polymer electrolytes. The impedance function can be expressed by

$$Z_{HN}(\omega) = \frac{R}{\left(1 + (j\omega\tau)^n\right)^p} \tag{2.39}$$

It represents a generalized RQ behavior, introducing an additional exponent p ($0 < p < 1$). The latter results in a non-symmetric impedance for $p \neq 1$. The influence of p on the impedance spectrum and the DRT is depicted in Figure 2.20. The decline of the impedance at higher frequencies becomes flatter for smaller p values, assuming the shape of a Gerischer impedance [94] for $p = 0.5$ and $n = 1$. This change has effect on the shape of the corresponding DRT. Additional side peaks occur at higher frequencies and describe a decreasing peak sequence. The reduction of n flattens the entire impedance curve and broadens the peaks in the corresponding DRT, similar to the effect of n_{RQ} for RQ-elements.

2.4.1.4 Warburg-element

The Warburg-element contains a whole range of impedance elements describing diffusion processes in electrochemical systems. The starting point for their derivation is Ficks first and second law:

$$j = -D\nabla c \qquad \frac{\delta c}{dt} = D\nabla^2 c \tag{2.40}$$

Here, D is the diffusion coefficient [$\text{cm}\,\text{s}^{-2}$], j the diffusion flux [$\text{mol/cm}^2\text{s}$] and c the concentration of the species [mol/cm^3]. The gradient operator ∇ represents the first

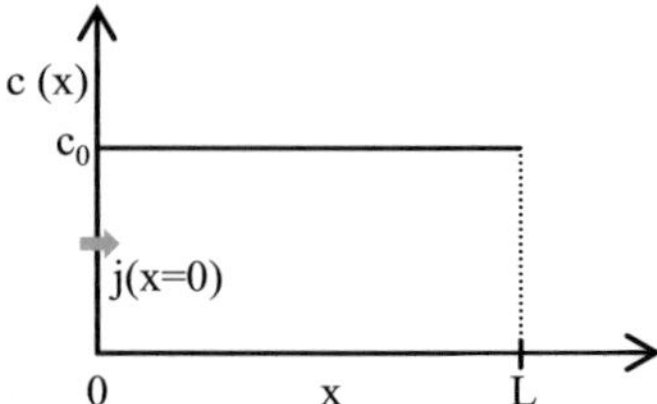

Figure 2.21: Concentration profile of the 1D diffusion zone at $t = 0$.

derivative with respect to the position. Assuming a one-dimensional diffusion path simplifies these equations and yields

$$j(t,x) = -D \cdot \frac{\delta c(t,x)}{\delta x} \qquad \frac{\delta c(t,x)}{\delta t} = D \cdot \frac{\delta^2 c(t,x)}{\delta x^2} \qquad (2.41)$$

These equations are transformed by Laplace transformation and solved by applying suitable boundary conditions. Following this, the function of concentration is entered into the Nernst equation in order to obtain a concentration dependent overpotential. This is then linearized and used for the calculation of the diffusion impedance from current flux $j(t,0)$ and the resulting overpotential. The derivation is explained in Refs. [53, 54, 95] in more detail. The applied boundary conditions and the respective impedance functions will be introduced in the following section in order to clarify the physical origin of different Warburg elements.

One-Dimensional Diffusion   A sketch of a one-dimensional diffusion zone is provided in Figure 2.21. The position $x = 0$ represents the interface between electrode and electrolyte where the lithium is inserted into the electrode. The position $x = L$ constitutes the end of the diffusion zone. First, it is necessary to define boundary conditions for both sides of the diffusion zone.

Generally, the first boundary condition at $x = 0$ is given by

$$j(t,0) = -D \cdot \frac{\delta c(t,x)}{\delta x}\Big|_{x=0} \qquad (2.42)$$

with a defined current flux $j(t,0)$ into the electrode. The second boundary condition depends on the kind of diffusion which is intended to be described. The first case describes a semi-infinite diffusion. In this case, the second boundary condition is defined at $x = \infty$ by $\frac{\delta c(t,x)}{\delta t}\big|_{x=\infty} = 0$. This yields the same impedance function as the previously introduced CPE element with an exponent $n_{CPE} = 0.5$.
Alternatively, two types of finite diffusion elements can be derived. These differ from each other due to the second boundary condition at $x = L$. The first finite diffusion

element is denominated as *Finite Space Warburg* element [96] (here and in the following text abbreviated as *FSW*, also published as *Finite Length Warburg* with *Open Circuit Terminus* [97] or *impermeable boundary diffusion* [54]). The corresponding boundary condition is defined by Equation 2.43:

$$j(t,x)|_{x=L} = -D \cdot \frac{\delta c(t,x)}{\delta x}|_{x=L} = 0 \tag{2.43}$$

This means that no flux is allowed at the second diffusion boundary $x = L$. For lithium-ion cells' active material this is comparable to the particle center where no flux of species is possible. Applying this boundary condition for the derivation of a *FSW* as described above leads to the following impedance function:

$$Z_{FSW}(\omega) = \frac{R_g T}{n F c_0 D} \cdot \frac{coth\left(\sqrt{\frac{j\omega}{D}} \cdot L\right)}{\sqrt{\frac{j\omega}{D}} \cdot L} \tag{2.44}$$

Here, R_g is the universal gas constant, T the temperature in Kelvin, n the number of exchanged charge carriers, F the Faraday constant, c_0 the concentration, D the characteristic diffusion coefficient and L the diffusion length. This impedance function results from the derivation assuming a one-dimensional diffusion path. It exhibits a characteristic 45° slope for high and pure capacitive behavior for low frequencies (see Figure 2.22c).

The implementation of $Z_{FSW}(\omega)$ can be simplified in order to obtain a more intuitive impedance function:

$$Z_{GFSW}(\omega) = R_w \cdot \frac{coth\left((j\omega\tau_w)^{n_w}\right)}{(j\omega\tau_w)^{n_w}} \tag{2.45}$$

Here, an additional exponent n_W ($0 < n_W < 0.5$) is introduced in order to represent the *Generalized Finite Space Warburg* element. n_W reduces the slope of the capacitive branch for decreasing values. The other model parameters are polarization resistance R_W and time constant τ_W. According to Equations 2.45 and 2.44 it is possible to express the time constant τ_W as $\tau_W = \frac{L^2}{D}$. It is therefore directly related to the diffusion length L and the diffusion coefficient D of the diffusing species.

The *Finite Length Warburg* element [96] is the second finite diffusion element (here and in the following text abbreviated as *FLW*, also published as *Finite Length Warburg* with *Short Circuit Terminus* [97] or *Nernstian diffusion layer* [54]). Its boundary condition is given by Equation 2.46:

$$\frac{\delta c(t,x)}{\delta t}|_{x=L} = D \cdot \frac{\delta^2 c(t,x)}{\delta x^2}|_{x=L} = 0 \tag{2.46}$$

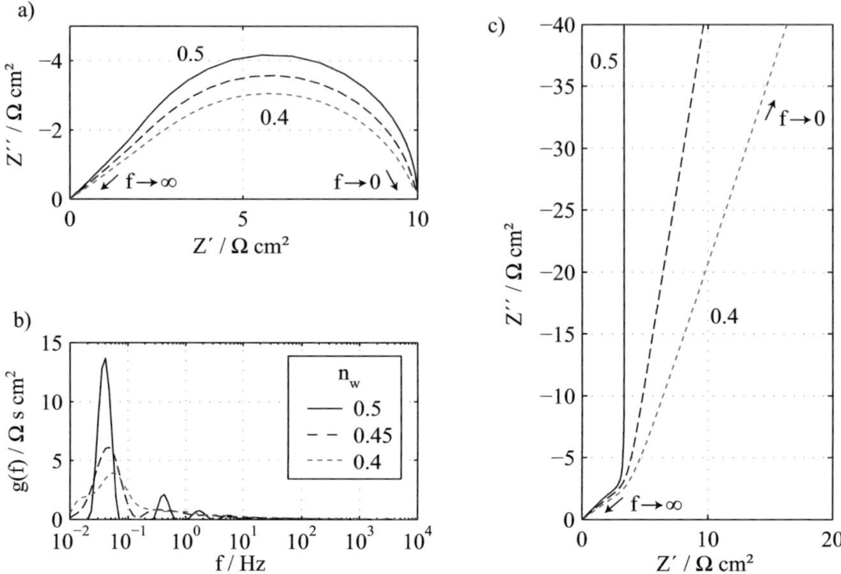

Figure 2.22: a) Nyquist plot and b) DRT of a simulated FLW impedance element($R_w = 10\Omega cm^2 \; \tau_w = 10s$) c) Nyquist plot of a simulated FSW impedance element($R_w = 10\Omega cm^2 \; \tau_w = 10s$).

This corresponds to a diffusion process through (not into) the diffusion zone. For this reason, not the diffusion zone volume, but rather its length is the limiting factor. The impedance function for the respective equivalent circuit element is expressed by

$$Z_{FLW}(\omega) = \frac{R_g T}{nFc_0 D} \cdot \frac{tanh\left(\sqrt{\frac{j\omega}{D}} \cdot L\right)}{\sqrt{\frac{j\omega}{D}} \cdot L} \tag{2.47}$$

for a one-dimensional diffusion path. It exhibits a characteristic 45° slope for high frequencies as well. However, for low frequencies it converges towards the real axis (see Figure 2.22a). This allows the calculation of the corresponding DRTs (Figure 2.22b). A characteristic peak sequence occurs in addition to the main peak due to the asymmetrical impedance of the FLW. The side peaks occur repetitively for higher frequencies with a decreasing height. It is again possible to rearrange Equation 2.47 to obtain a more intuitive impedance function:

$$Z_{GFLW}(\omega) = R_w \cdot \frac{tanh\left((j\omega\tau_w)^{n_w}\right)}{(j\omega\tau_w)^{n_w}} \tag{2.48}$$

Here, an additional exponent n_w ($0 < n_w < 0.5$) is again introduced in order to represent the *Generalized Finite Length Warburg* element. n_w flattens the impedance

Figure 2.23: Geometries for the derivation of 1D-, 2D- and 3D-diffusion elements.

spectrum as it was already described for RQ-elements. The other model parameters are polarization resistance R_w and time constant τ_w.

Two-Dimensional Diffusion The previously introduced Warburg elements assume a diffusion process along one-dimensional diffusion paths. This assumption can be justified for some electrode materials in lithium-ion cells, for instance $LiFePO_4$ or $Li_4Ti_5O_{12}$ (see Section 2.1.2). Yet there are several electrode materials allowing two- or three-dimensional diffusion. Therefore it is necessary to derive FLW and FSW for 2D and 3D diffusion as well. The derivations are based on symmetrical geometries that adequately represent the diffusion zone.

A cylindrical geometry is used for 2D-diffusion, describing a diffusion along the radial axis (Figure 2.23). Consequently, cylindrical coordinates are used for the derivation of the impedance function [54]. Applying the FSW boundary conditions at the electrode surface (Equation 2.42) and in the center of the cylinders' volume (Equation 2.43) and deriving the generalized impedance function leads to the following equation:

$$Z_{GFSW,2D}(\omega) = R_w \cdot \frac{I_0\left((j\omega\tau_w)^{n_w}\right)}{(j\omega\tau_w)^{n_w}\, I_1\left((j\omega\tau_w)^{n_w}\right)} \tag{2.49}$$

Here, R_w, τ_w and n_w are the same Warburg parameters as previously introduced. I_0 and I_1 are modified zero and first-order Bessel functions of the first kind [54, 98]. As these functions diverge for large arguments ($I_{0/1} \xrightarrow{x \to \infty} \infty$) it is necessary to approximate the impedance function for the calculation at high frequencies. An approximation is possible by using the following equation:

$$Z_{GFSW,2D,HF}(\omega) = R_w \cdot \frac{1 + \frac{1}{8(j\omega\tau_w)^{n_w}}}{(j\omega\tau_w)^{n_w} \cdot \left(1 - \frac{3}{8(j\omega\tau_w)^{n_w}}\right)} \tag{2.50}$$

The implementation of the two-dimensional FSW element for the evaluation of impedance spectra comprises of both Equations 2.49 and 2.50 in order to allow for stable numerical

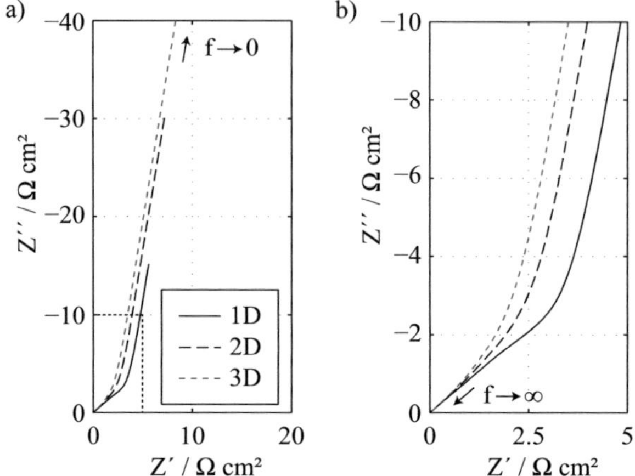

Figure 2.24: a) Nyquist plot of simulated 1D-, 2D- and 3D-FSW impedance element ($R_w = 10\,\Omega\mathrm{cm}^2$, $\tau_w = 10\,\mathrm{s}$, $n_w = 0.45$). b) Magnification of high frequency range.

calculation during the fitting procedure. The latter is guaranteed by checking the maximal frequency processable for the simulation and by approximating the impedance function by Equation 2.50 for all frequencies above.

Likewise, it is possible to apply the cylindrical geometry for the derivation of a two-dimensional FLW. However for that it is necessary to define two additional parameters for the description of the diffusion process' geometry. 1This is not introduced in this thesis as it is not needed for the following evaluations.

Three-Dimensional Diffusion   The Warburg element to describe a three-dimensional diffusion can be derived by applying a spherical geometry with three-dimensional diffusion paths (see Figure 2.23). The derivation of 3D FSW follows the steps previously explained, using a spherical coordinate system and the boundary conditions for finite space behavior [54]. The respective impedance function is defined by

$$Z_{GFSW,3D}\,(\omega) = R_w \cdot \frac{tanh\left((j\omega\tau_w)^{n_w}\right)}{tanh\left((j\omega\tau_w)^{n_w}\right) - (j\omega\tau_w)^{n_w}}. \tag{2.51}$$

The fundamental shapes of 1D-, 2D- and 3D-FSW elements' impedance spectra are similar. Each of them exhibits a characteristic 45° slope for high and capacitive behavior for low frequencies. The main difference among them is the transition between the high frequency and the low frequency regions, illustrated in Figure 2.24. The bend between the two characteristic slopes is very distinct for the 1D FSW and smooths out for 2D- and 3D-diffusion. Furthermore, the capacitive behavior is more pronounced for the higher dimensional diffusion elements.

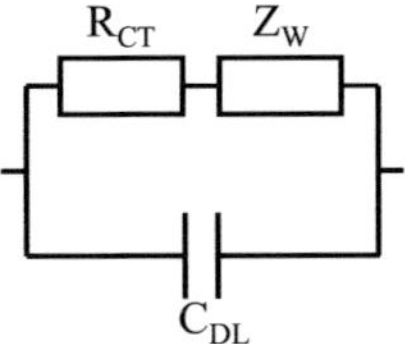

Figure 2.25: Structure of the commonly known Randles circuit [53, 102].

2.4.2 Randles Circuit

The impedance models previously introduced are able to represent specific loss mechanisms separately. RC-elements were proposed to describe charge transfer processes, whereas Warburg elements were presented as diffusion elements. However, in real systems a separate examination of charge transfer and diffusion at the same interface is not possible. Randles et al. [99] established an impedance model structure representing the combination of both mechanisms in one impedance element, the so-called Randles circuit [53, 99–102]. It consists of a serial connection of charge transfer resistance and diffusion element in parallel with the double layer capacitor as illustrated in Figure 2.25. This concept is based on the assumption that the charge transfer overpotential is directly connected with the solid state diffusion and cannot be considered as an independent loss process. Accordingly, it is not possible to uncouple the dynamic behavior of charge transfer from diffusion by an independent RC-element.

The effect of such a model structure on the impedance spectrum in practical use is illustrated in Figure 2.26. Simulations of an RC-element and a one-dimensional FSW connected in serial and by the Randles circuit are compared for different diffusion time constants. The gap between FSW and RC time constant varies from two to four decades. The comparison reveals that, for low frequencies, the model structure has a significant impact in none of these three cases. However, in the high frequency region the gap between RC- and FSW time constant is crucial. The impact is clear in the case of a gap of two decades as can be seen in Figure 2.26a. In this case, the bend separating the capacitive branch and the semicircle is not very pronounced and is further reduced when using a Randles circuit. For larger gaps, no strong influence occurs (Figure 2.26b and c). These results show that the need of a Randles circuit is limited to charge transfer and diffusion processes with similar time constants. For such parameters the use of it can be advantageous. Otherwise it does not have an effect.

2.4.3 Transmission Line Models

Transmission Line Models (TLM) define a complex ECM design in order to take the impact of the electrode structure on its loss mechanisms into account. To be exact, TLMs are designed to describe porous electrode structures comprising an electron and an ion conducting path and an interface between them. The origin of an interaction

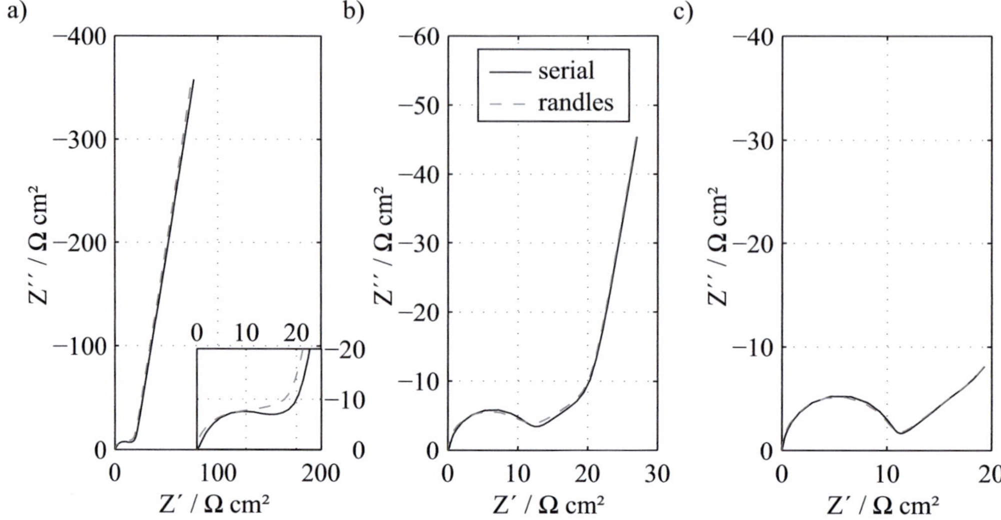

Figure 2.26: Nyquist plot of a simulated RC-element connected with a 1D FSW in the Randles structure for varying time constant τ_{FSW}: a) τ_{FSW} =1 s, b) τ_{FSW} =10 s and c) τ_{FSW} =100 s (R_{RC} =10 Ωcm^2, τ_{RC} =0.01 s, R_{FSW} =30 Ωcm^2, n_{FSW} = 0.45).

amongst them is a limited electronic or ionic conductivity, leading to potential gradients along the electrode layer thickness. As a consequence the impedance spectrum of the electrode is modified. These effects can be represented by a TLM as illustrated in Figure 2.27.

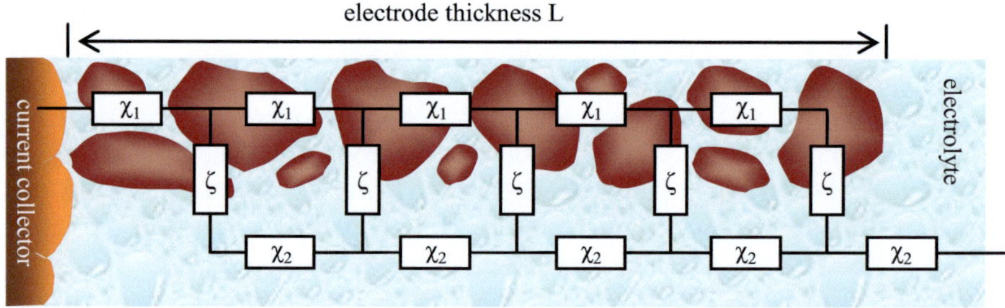

Figure 2.27: Structure of transmission line models for porous electrodes.

The equivalent circuit element ζ to represent the interface between electrode and electrolyte (including charge transfer, reaction layers, solid state diffusion) is integrated into the TLM structure together with the electronic path χ_1 through the electrode and the ionic path χ_2 through the electrolyte in the pores. The latter comprises of distributed resistance elements defined by

$$\chi_{1/2} = r_{1/2} = \frac{1}{A\sigma_{e^-/ion}}[\Omega cm^{-1}] \tag{2.52}$$

using a specific conductivity $\sigma_{e^-/ion}$ and an electrode area A. In some cases, $r_{1,2}$ may contain polarization losses instead of pure ionic or electronic conductivity. The polarization resistance of both paths over the entire electrode layer can be calculated from Equation 2.53:

$$R_{1/2} = r_{1/2} \cdot L[\Omega] \tag{2.53}$$

Here L defines the thickness of the electrode. In contrast, the specific interface resistance ζ results from the overall interface impedance Z_i by

$$\zeta = z_i = Z_i \cdot L[\Omega cm]. \tag{2.54}$$

The overall impedance of the TLM can be expressed by the following equation:

$$Z_{TLM,2path} = \frac{\chi_1\chi_2}{\chi_1 + \chi_2}\left(L + \frac{2\kappa}{sinh\left(\frac{L}{\kappa}\right)}\right) + \kappa\frac{\chi_1^2 + \chi_2^2}{\chi_1 + \chi_2}coth\left(\frac{L}{\kappa}\right) \tag{2.55}$$

with

$$\kappa = \left[\frac{\zeta}{\chi_1 + \chi_2}\right]^{\frac{1}{2}}. \tag{2.56}$$

Here and in the following thesis, this kind of TLM will be denominated as two path TLM, as the electronic and the ionic paths contribute to the impedance spectrum. However, their contribution depends on the relation between electronic and ionic conductivity and the interface resistance. In the case that one of the conductivities is much higher, the corresponding path can be considered as a short circuit. If for instance $\chi_1 = 0$ can be assumed, Equation 2.55 is reduced to

$$Z_{TLM,1path} = (\zeta\chi_2)^{\frac{1}{2}} coth\left[(\chi_2/\zeta)^{\frac{1}{2}}\right]. \tag{2.57}$$

Accordingly, this TLM will be denominated as a one path TLM.
A meaningful application of TLMs depends on the knowledge of several physical parameters. The electrode layer thickness L is easy to measure and the interface resistance ζ is obtained by fitting. Choosing a reasonable $\chi_{1/2}$ is more challenging. It is possible to measure the electronic layer conductivity of an electrode directly [103]. However,

the direct measurement of ionic conductivity in the pores and the resulting χ_2 is not possible. χ_2 can be determined by measuring the specific ionic conductivity of the electrolyte and by determining the microstructural parameters of the electrode. With these values and the electrode thickness L it is possible to calculate the expected χ_2. The entire procedure is demonstrated in Section 7.5 for real electrode structures.

The effect of a TLM structure on impedance spectra is demonstrated by the simulations depicted in Figure 2.28. One simple RC-element is used as ζ in a one path TLM with short circuited electronic path ($\chi_1 = 0$). The ionic conductivity and therefore $\chi_2 = r_2 = \frac{R_2}{L}$ is varied in order to demonstrate its impact. For the smallest value of $R_2 = 1\,\Omega\mathrm{cm}^2$ the spectrum represents an almost ideal RC-element. Only a small $45°$ slope can be observed for high frequencies. The DRT in Figure 2.28b shows one peak at the characteristic frequency of the RC-element and a small side peak at high frequencies due to the $45°$ slope. The impedance spectrum is modified for increasing R_2. The $45°$ slope and also the overall polarization increases significantly. In the DRT this is represented by an increasing sequence of side peaks. Finally, the largest R_2 causes a much larger polarization and a smooth drop of the impedance spectrum at higher frequencies. This results in a significant broadening of the corresponding DRT to the high frequency region. Not the main peak but rather the area under the entire DRT increases strongly.

These theoretical considerations are important for a meaningful evaluation of impedance spectra and DRTs. The simple assumption that one peak corresponds to one electrochemical process does not hold for complex electrode structures. The interaction between the electrode loss processes and the electrode structure hinders this simple interpretation. A knowledge of electrode parameters such as thickness, porosity or conductivity is necessary in order to estimate the microstructural influence. Having this knowledge allows for more physically based impedance modelling. The derivation of TLM structure was introduced by Euler and Nonnenmacher in 1960 [104] and further developed by Levie et al. [105].

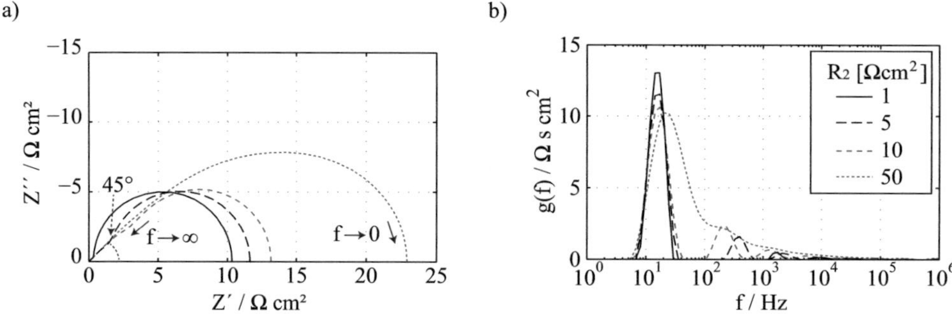

Figure 2.28: a) Nyquist plot and b) DRT of one path TLM for varying ionic conductivity using one RC-element as interface process ζ ($R_{RC} = 10\,\Omega\mathrm{cm}^2$, $\tau_{RC} = 10\,\mathrm{ms}$, $L = 48\,\mathrm{\mu m}$, $R_2 = R_{ion} = 1\,\Omega\mathrm{cm}^2$ to $50\,\Omega\mathrm{cm}^2$, $\lambda_{DRT} = 10^{-6}$).

3 State Of The Art

This chapter introduces the common impedance analysis known from the literature to investigate lithium-ion cells and electrodes. It starts with the analysis of experimental cells and single electrodes via impedance spectroscopy and equivalent circuit modelling. Several studies will be introduced according to their impedance modelling approach. The results of these studies will be compared later to the results obtained in this thesis in the corresponding discussion chapters. Finally, this chapter concludes with a survey on the electrochemical analysis of commercially available lithium-ion cells.

3.1 Equivalent Circuit Modelling of Lithium-Ion Electrodes

Electrochemical Impedance Spectroscopy was introduced for the first time for the investigation of liquid electrolyte systems by Sluyters et al. in 1960 [106]. It was established for the analysis of energy conversion systems in 1987 after Ref. [107] was published by Macdonald et al. In this book, the fundamentals of EIS and equivalent circuit modelling were presented. All kinds of equivalent circuit elements such as RC- or RQ-elements and model structures such as transmission line models or Randles circuits have been presented and motivated theoretically. Even the problem of ambiguous equivalent circuit models and difficult separation of processes with closed time constants was already recognized. To this day, the most common approach to evaluate impedance spectra of lithium-ion cells is the introduction of an equivalent circuit model according to the Nyquist plot of the impedance spectrum and a subsequent CNLS-fitting. The fit accuracy is then used as quality criterion to judge the impedance model proposed and the fit parameters obtained.

This basic procedure to analyze lithium-ion batteries and lithium-ion battery electrodes in experimental cells is well-established. It is applied to identify loss mechanisms in electrodes and to understand the performance-limiting factors in the latter. The frequency range usually covered by EIS includes frequencies between $100\,\mathrm{kHz}$ and $10\,\mathrm{mHz}$ [28, 108–111] or, in some rare cases, down to the mHz-range [112]. Several methods are used to further analyze the spectra obtained.

First, as introduced above, an equivalent circuit model is chosen *a priori* according to the shape of the measured impedance spectrum. It usually contains a serial ohmic resistance, a Warburg diffusion element and several RQ- or RC-elements, depending on the number of semicircles identified in the spectrum [43, 44, 111]. For more complex impedance spectra exhibiting very flat semicircles it is also common to use a serial connection of a higher number of RC-elements in order to describe the impedance spectrum appropriately without having a hint on the correct number [59]. This leads to a good fit quality but to very ambiguous fit results.

The use of more complex model structures like transmission line models, known to describe the complex microstructure of porous electrodes better, is not common as the parameters needed for a meaningful parametrization are not available. Two examples for the application of TLMs for lithium-ion battery electrodes have been presented in Refs. [102, 113]. However, the microstructure parameters necessary for an estimation of meaningful parameters are in these studies not available as well.

The proposed models are subsequently used to determine the dependencies of each loss process on SOC [114–116] or temperature [43, 44] and to understand the physical origin of each contribution.

Another way to analyze the impedance spectra and to identify the physical origin of the occurring loss processes is given by a variation of the electrode structure. Several parameters such as layer thickness [102], layer composition [28, 109, 112], electrode pressure [117] or electrolyte [43, 44] are varied and the impact on the impedance spectrum is observed. The effect of these steps is evaluated qualitatively in order obtain a meaningful interpretation. However, the direct correlation between microstructure and impedance spectra was not possible until now.

The impedance spectrum for frequencies below 1 mHz is usually not measured. In the literature, GITT and PITT measurements are applied in order to investigate the low frequency loss processes such as solid state diffusion [77, 110, 118, 119]. However, these methods assume the solid state diffusion to be dominant in this frequency range and do not analyze the corresponding frequency range without presumptions.

The DRT provides an advanced method to evaluate impedance spectra which has already been used for SOFC analysis [70] and has recently been applied for the time-domain simulation of lithium-ion batteries [120]. In this thesis, the DRT approach is applied for the first time as a starting point for the analysis of lithium-ion cell electrodes. For that, equivalent circuit models are developed based on the previous DRT-analysis in order to determine impedance models without presumptions.

Simple impedance models have been developed for lab-scale electrodes, whereas complex TLM models have been designed for commercially available electrodes. The models are based on DRT analysis and complemented by microstructure parameters obtained from 3D microstructure reconstructions developed in Ref. [1]. Additionally, the standard impedance spectroscopy was complemented by time domain measurements introduced in Ref. [74] in order to determine the low frequency impedance spectrum of lithium-ion cell electrodes for the first time down to the µHz region. In each chapter, the results obtained by this new approach will be compared in detail to the results obtained by the conventional approach described above.

3.2 Electrochemical Analysis of Lithium-Ion Cells

In general, there are two fundamentally different goals of the electrochemical investigation of lithium-ion full cells. First, the determination of external parameters such as the impedance spectrum is needed in order to predict the cell voltage and to estimate the SOC or the SOH. These properties can be obtained from external full cell impedance or time domain measurements without a detailed insight into the cell [74, 121–123]. The second possible goal is the obtaining of a deeper understanding of the electrochemical processes taking place in the cell during operation. Such knowledge is crucial for the improvement of cells and for the identification and interpretation of aging mechanisms. Furthermore, this information can be integrated into the applications introduced above. The first step to obtaining a rough physical interpretation of a lithium-ion cell's impedance spectrum is the separation of anode and cathode losses. One way to achieve this separation would be the introduction of a reference electrode into the cell case. This can be done if the cells are assembled by the researcher and if there is surplus space inside the cell case for the additional electrode. Such a cell setup is very helpful due to the *in-situ* separation of the electrode contributions. However, it is not trivial to fabricate such cells stably and with a reference electrode suitable for impedance spectroscopy. Studies presented in 2000 [124, 125] and 2005 [126] were the first to have a corresponding cell setup; however, the reference electrodes applied were not suitable or had suitability constrictions for impedance measurements. New publications of these groups continuing that research are not available. One general drawback of this approach is that it cannot be pursued for commercially available cells which have to be investigated as delivered without having a reference electrode.

In that case, opening the cell is the only way to separate the contributions of its single components. There are several studies published by Abraham et al. [127–130] that demonstrate the opening of new and aged high-power 18650 cells with NCA/graphite electrodes. After opening, the electrodes are harvested and analyzed using several morphological and analytical tools such as SEM, XRD or Raman spectroscopy. Furthermore, electrochemical measurements such as pulse experiments or impedance spectroscopy were performed in experimental cell housings. A reference electrode was applied in order to separate the contributions of each electrode. The impedance spectra are analyzed qualitatively by comparing the Nyquist plots of both electrodes and assigning the main aging to the cathode. A detailed analysis and interpretation of the impedance spectra was not conducted.

The next study, dealing with the opening of commercially available cells and the subsequent analysis of its components, was published 2013 as a part of this thesis in Ref. [60]. In parallel, Refs. [131, 132] were published in order to determine the major contributions to the aging of commercially available lithium-ion cells. A detailed analysis of the impedance spectra such as a physical interpretation or the development of impedance models is not provided. In Ref. [131], a $LiFePO_4$/graphite cell was opened and the electrodes were analyzed by chemical and morghological methods such as SEM, XRD or FTIR. Moreover, the electrodes were analyzed electrochemically in experimental test cells by charge/discharge, OCV and impedance measurements. The impedance spectra were evaluated qualitatively just as introduced above. Therefore, the Nyquist plots of both electrodes were compared in order to give a rough estimation of both electrodes' contributions. This was done without giving a physical interpretation for the characteristics in the impedance spectrum.

Another study was recently published by Zavalis et al. [133]. Here, the electrode contributions in a $LiFePO_4$/graphite cell were separated by performing impedance spectroscopy in experimental cells via reference electrode. The results were compared with a one-dimensional electrochemical model previously developed in order to assign the aging to certain mechanisms.

In summary, it can be stated that recently the detailed analysis of commercially available cells and its components attracts more attention as the understanding of the relevant loss mechanisms is a crucial factor to identify degradation mechanisms. The methods usually applied for this purpose are a combination of standard impedance spectroscopy and other chemical analysis.

In this thesis, a new approach is introduced to investigate commercially available lithium-ion cells and its extracted electrodes via DRT analysis in order to obtain more information from impedance measurement evaluations. Furthermore, the standard impedance spectroscopy is complemented with time domain measurements to extend the frequency range. Finally, this procedure is combined with comprehensive impedance modelling of single electrodes in order to better understand the physical origin of the commercial full cells' impedance spectra.

4 Experimental

In this chapter, the experimental setup used in the present thesis is introduced. It starts with cell housings and configurations, continues with the opening of commercially available cells and concludes with the measurement equipment applied and the measurement data quality achieved. The detailed composition and measurement procedure for the cells measured in this thesis will be described at the beginning of each chapter.

4.1 Experimental Cell Housings

The investigation of new battery materials is usually done in experimental cells at lithium-ion battery research. Such cells are advantageous because they are easy to assemble, need a small amount of material due to a small electrode area (in the range of a few cm^2) and enable the assembly of any kind of cell configuration combining different electrodes (introduced in Section 4.2). They allow for an analysis of electrodes without the influence of large cell geometries or temperature gradients. All experimental cells used in this thesis were assembled in a glovebox (MBraun, Unilab and MB200MOD) in a pure argon atmosphere in order to prevent reactions between lithium, electrodes and electrolyte with oxygen, water or other contaminating species. Such contamination would lead to a strong degradation of the cells and prevent reproducible measurement results. In the literature, several types of experimental cell housing serve as a basis for experimental cell measurements. Coin cells are very popular because they are an industry standard, providing reliable measurements concerning cell stability and a good sealing. However, these cell housings cannot be re-opened without destroying the electrodes and do not allow for the simple insertion of a third electrode as reference electrode. Another widespread kind of experimental cell housing which can be used in three electrode configuration with an additional reference electrode is integrated into Swagelok parts. However, this cell housing was found not to be suitable for impedance spectroscopy via reference electrode [134]. Furthermore, the assembly and the sealing of this cell type turned out to be complicated and not satisfactorily reproducible.
A new cell setup was developed by EL-Cell [135], presented in Figure 4.1. It is convenient and reproducible concerning the cell assembly and provides a reliable sealing concept. The housing provides a reference electrode connection and is relatively flexible for the use of alternative reference electrode geometries which will be introduced in Section 4.2.4. All cells measured in this thesis were assembled in such EL-Cell housing.
Recently, another cell housing was developed at IWE, optimized for the integration of the newly developed reference electrode geometry.

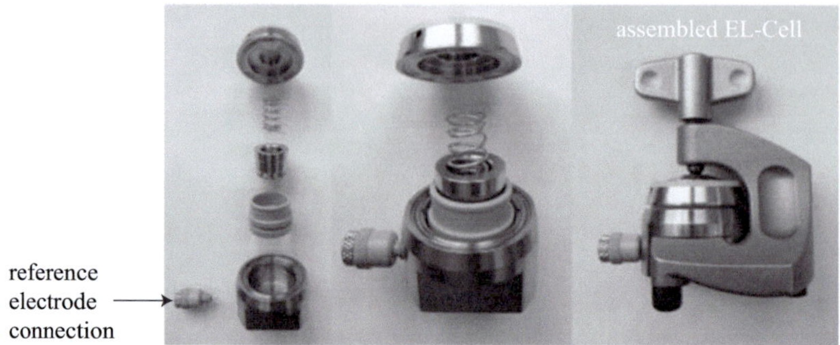

Figure 4.1: EL-Cell housing with integrated reference electrode connection [135].

4.2 Experimental Cell Configurations

Experimental cells enable various combinations of different electrodes as illustrated in Figure 4.2. Cathode and anode are combined in full cells in order to investigate electrodes in the same configuration as in standard lithium-ion cells. Furthermore, it is possible to assemble half-cells in which cathode or anode are combined with lithium metal as the counter electrode. Another configuration is given by symmetrical cells containing two cathodes or two anodes, respectively. Applications, advantages and disadvantages of each cell configuration will be introduced in the following sections. Every cell configuration can be complemented by an additional reference electrode in order to separate the contributions of the electrodes.

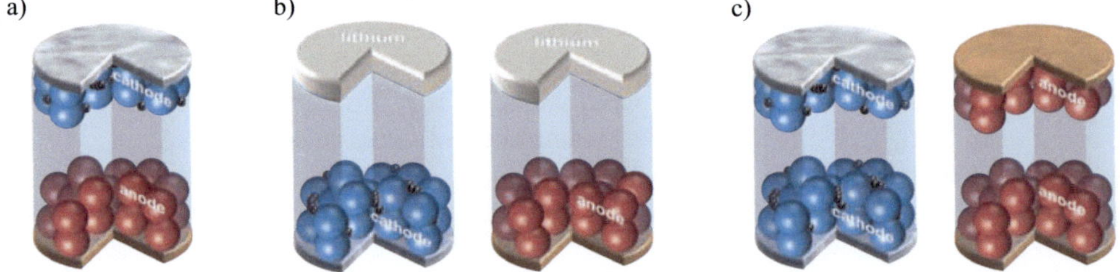

Figure 4.2: Experimental cell configurations: a) Full cell, b) half-cells and c) symmetrical cells.

4.2.1 Half-Cells

The most popular cell configuration for material research is the half-cell, combining the electrode under investigation with a lithium metal anode. This allows for an undistorted determination of the electrode capacity because a surplus of lithium is available due to the thick lithium metal anode. Therefore, the cell capacity is limited by the electrode investigated and also a lithium consumption during cycling does not affect the surplus lithium reservoir. Half-cells are also used for the measurement of characteristic OCP-curves because of the constant potential of lithium metal (per definition $0\,\text{V}$).

Half-cells also exhibit several advantages concerning the impedance analysis. First, the measurement of impedance spectra is possible over the entire SOC range of the electrode under investigation. Moreover, lithium is deposited directly on the surface of the lithium metal anodes and therefore does not need a solid state diffusion transport into the active material. This means that no low frequency loss processes as well as no capacitive behavior are expected for the impedance spectrum of lithium metal anodes. From theoretical point of view, it is also advantageous to use half-cells as, in this setup, the counter electrode is always similar and therefore produces a well-known contribution to the impedance spectrum.

One disadvantage of lithium metal anodes is their strong reactivity with the organic electrolyte which leads to electrolyte degradation and thus a possible influence on the cells' stability. Similarly, an excessive SEI formation is expected due to the continuous dissolution and deposition of lithium at the lithium metal surface, leading to a degradation of the electrolyte and the impedance spectrum of the lithium metal. The real contribution of lithium metal anodes to the impedance spectrum and the applicability of half-cell measurements for impedance analysis will be discussed in Chapter 5.

4.2.2 Symmetrical Cells

Symmetrical cells are assembled from two equal cathodes or anodes and are therefore the tool for a reliable impedance analysis of one type of electrode without the interaction or overlap with other electrodes [136, 137]. However, the electrodes under investigation have to be charged in a half-cell to the desired SOC previous to the assembly of symmetrical cells. This is an additional step which makes the use of symmetrical cells relatively time-consuming and error-prone. Furthermore, the latter cannot be varied in SOC as charging changes both electrodes asymmetrically.

Recently, symmetrical cells have been used for the investigation of electrode degradation in experimental cells [138]. In this thesis, symmetrical cells are used for a separation of electrode contributions in half-cells' impedance spectra.

4.2.3 Full Cells

The experimental cell configuration which is most similar to the real application is the combination of cathode and anode in a full cell. It enables the investigation of the electrodes as they are used in standard lithium-ion cells. In this thesis, experimental full cells are used for the impedance analysis of electrodes harvested from commercially available cells. This configuration provides the advantage of comparable exploitation of the electrode capacities and almost similar SOCs compared to the commercial cell.

Furthermore, no lithium metal is available in this cell configuration, leading to a decreasing electrolyte degradation. At the same time, this is a disadvantage of this cell configuration because the limited amount of lithium leads to capacity fading due to the SEI formation. Another drawback is the overlap of anode and cathode losses in the impedance spectrum of full cells. However, this problem can be solved by introducing a reference electrode which will be presented in the next section.

4.2.4 Reference Electrode

The concept of a reference electrode is based on the introduction of a third electrode into the cell. Therefore it is also called three-electrode-setup. The system is stimulated by a working and a counter electrode, whereas the reference electrode is free of any current flow. The latter is used to measure the fixed potential at one position in the electrolyte. Relative to this fixed potential it is possible to determine the overpotentials of the working and the counter electrode (corresponding to cathode and anode) separately. It can be used for charge discharge experiments as well as for impedance spectroscopy. However, a reliable separation of both electrodes' overpotentials necessitates an ideal position of the reference electrode and a perfect cell setup [134, 139, 140].
Three different reference electrode setups have been analyzed against the applicability for impedance spectroscopy in a study project by Amine Touzi [141]. All setups analyzed are illustrated in Figure 4.3: a point-like reference electrode provided as a standard in the EL-cell housing [135], a wire-like and a mesh-like geometry. All possibilities have been investigated by FEM simulations and by experimental cell measurements. It turned out that the standard point-like reference electrode, which is usually realized by a piece of lithium metal in the EL-cell housing, leads to artifacts in the impedance spectrum due to geometrical and electrical asymmetries. This result was already presented by Moses Ender in Ref. [134] for a comparable reference electrode setup in Swagelok-type cells. The use of a wire-like reference electrode turned out to be less sensitive to such asymmetry and the resulting artifacts. Lastly, a mesh-like geometry was simulated. This geometry has proven to be most insensitive to artifacts by asymmetries of the cell setup, allowing for a superior electrode separation in experimental cells. These simulations have also been validated by measurements, assembling the EL-cell housing with these kinds of reference electrodes.

The latter were realized by a piece of lithium metal in the case of a point-like reference electrode. Steel wire and steel mesh were applied for the other two geometries as lithium wire and lithium mesh cannot be produced. The measurements proved the simulation results; however, one drawback of this new reference electrode setup arose because steel is electrochemically not stable in an organic electrolyte. This led to a potential drift of the reference electrode during impedance measurements, decreasing measurement data quality. Checking the validity of the impedance spectra measured in this configuration is therefore necessary. It will be introduced in Section 4.5.
One solution in order to overcome this problem of potential drift is provided by coating the reference electrode mesh with active material. Two kinds of active material are

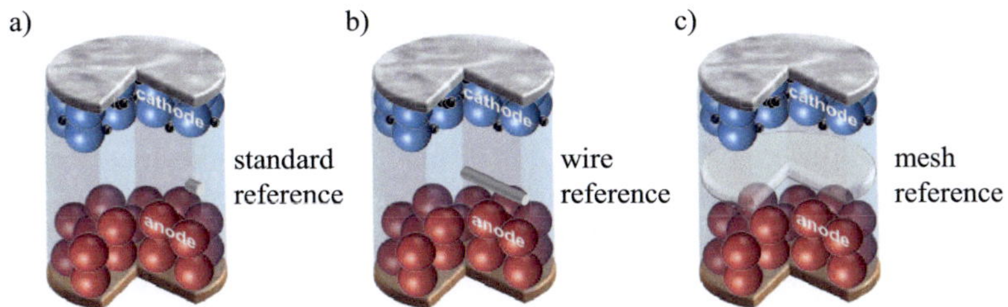

Figure 4.3: Different reference electrodes using a) point-, b) wire- and c) mesh-like geometry.

suitable for this kind of application: $LiFePO_4$ and $Li_4Ti_5O_{12}$ [142]. Both provide a stable potential over a large range of SOC. Figure 4.4 shows one of the uncoated steel meshes applied in this thesis (195 µm mesh size, 70 µm thickness) and a steel mesh coated by $Li_4Ti_5O_{12}$. A disadvantage of the $Li_4Ti_5O_{12}$-coated steel mesh is the need for lithium in order to lithiate the $Li_4Ti_5O_{12}$. However, this is not a critical point as the reference electrode capacity is negligibly small compared to the main electrodes. First tests have successfully been conducted on coated mesh-like reference electrodes; the measurements in this thesis are still conducted using an uncoated steel mesh.

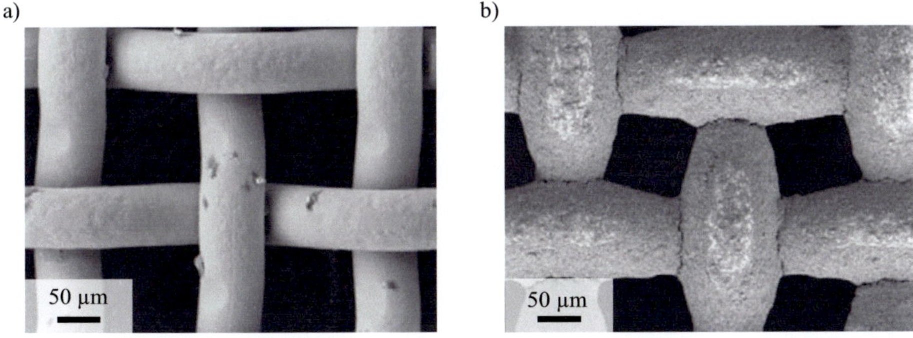

Figure 4.4: a) Uncoated and b) $Li_4Ti_5O_{12}$-coated steel mesh like it is used as reference electrode.

Experimental cells with such a reference electrode need two layers of separator in order to separate reference and working electrodes. Furthermore, they need a surplus of electrolyte to cover mesh and separator. This makes the setup more complicated to assemble and error-prone due to a possible drying out of the mesh. However, it is the best setup to analyze full cells via impedance spectroscopy with a simultaneous separation of electrode losses.

4.3 Opening Lithium-Ion Cells

In this thesis, commercially available lithium-ion cells (introduced in more detail in Appendix A) were opened in a glovebox in order to investigate the components in experimental cell configuration. This procedure can be applied to any kind of lithium-ion cell; however, the tools and methods for opening are different, depending on the type of cell to be opened. In this section, the opening of an 18650 cell is described. It comprises of several steps listed in the following enumeration:

1. Remove the film which covers the cell case.
2. Cut the top and the bottom of the cell with a pipe cutter.
3. Peel the metallic cell case from the electrode roll.
4. Unroll the electrode/separator stack.
5. Separate cathode, anode and separator.
6. Wash the electrodes in a solvent.
7. Remove the back side in case of double-side coated electrodes.
8. Punch out the electrodes for the experimental cells.
9. Wash the electrodes in a solvent and dry them subsequently.

A series of pictures in Appendix B illustrates all steps in detail. During the entire procedure, it is crucial to prevent short circuiting of the cell and, after dis-assembly, of the cathode and anode layers. This is indispensable in order to reduce the impact of the cell opening on the measurement results of its individual components. Another critical issue is the washing and the preparation of the electrodes after extracting them from the commercial cell case. Above all, the removal of the back side coating of the double-side coated electrodes is the critical step as the resulting mechanical stress also influences the coating on the back side. Several preparation methods have been compared in this thesis and will be introduced in Chapter 8.

4.4 Measurement Setup

All measurements presented in this thesis have been performed using a multichannel 1470E cell-test system (Solartron) with Multistat software (Scribner). The cell-test system consists of a galvanostat/potentiostat unit and a frequency response analyzer (FRA) to perform impedance spectroscopy. Furthermore, an Agilent 34970A voltmeter was used to increase the accuracy of the cell voltage during time domain measurements. All measurements were performed in climate chambers (Weiss WK1180, Vötsch VT4002) in order to guarantee a stable temperature and to analyze temperature dependencies. Impedance spectra were measured between 1 MHz and 10 mHz. The detailed settings will be introduced for each study in the corresponding chapter. The measurement data analysis was performed in Matlab and bases on KK-analysis with subsequent DRT-evaluation and CNLS-fitting.

4.5 Measurement Data Quality

The quality of the measured impedance data is of crucial importance for a meaningful analysis by DRT and equivalent circuit modelling. It depends on meeting the preconditions for impedance spectroscopy as already introduced in Section 2.2.2 and on a high signal data quality. The KK-residuals are a well-known mathematical measure to check the measurement data quality. Figure 4.5 presents the KK-residuals of four different cell configurations used in this thesis. All configurations show increasing residuals for low frequencies below 0.1 Hz. The strongest increase occurs for the measurement via reference electrode. This is caused by the unstable potential of the steel mesh reference electrode. This limits the reliability of reference electrode impedance spectra below 0.1 Hz.

Another significant increase of residuals can be observed for the full cell and the reference electrode measurements in Figures 4.5c and d at 20 kHz and above 100 kHz. The increase at 20 kHz is caused by an automatic switch of the measurement range and necessitates a careful evaluation of this frequency range. The increase above 100 kHz is caused by the internal compensation of the cable inductance by the measurement equipment. These frequencies are therefore not evaluated by DRT-analysis in this thesis.

Another obvious difference occurs among the half-cell and the symmetrical cell between 10 Hz and 1000 Hz. The half-cell shows higher residuals because a switch of the measurement range occurs at 10 Hz. However, another important reason is that this frequency range is dominated by the lithium anode (this will be shown later). The use of a lithium anode leads, in general, to higher residuals due to its time variant impedance spectra. Furthermore, this indicates non-linear contributions of the lithium anode which were identified in the study project of Bogdan Brad [143].

Except for a part of the frequency ranges described above, the residuals are below 0.4 % over the main part of frequencies for all cell configurations applied. This allows for a meaningful DRT and ECM-analysis. The limitations described above are taken into account during the evaluation.

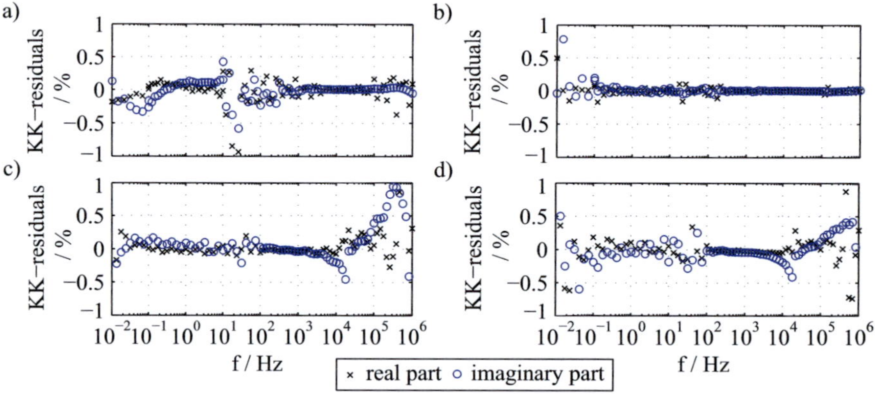

Figure 4.5: Kramers-Kronig residuals of a) half-cell, b) symmetrical cell, c) full cell and d) reference electrode measurement. In general, the focus of DRT evaluation of EIS is set on frequencies between $100\,\text{kHz}$ and $100\,\text{mHz}$ in the following chapters. Frequencies around a changing measurement range as i.e. at $20\,\text{kHz}$ are interpreted with a reduced emphasis on the DRT evaluation.

5 Modelling of Lab-Scale Cathodes

The goal of this chapter is the identification of electrochemical loss processes for lab-scale lithium-ion battery electrodes. Electrochemical impedance spectroscopy is applied as it is known to be suitable for the separation of loss mechanisms occurring at different time scales. The impedance spectra are afterwards analyzed by the DRT (Section 2.2.3) which is known from solid oxide fuel cells (SOFC) analysis [71] and from time domain simulations of lithium-ion batteries [74]. These investigations are conducted using experimental cells in different cell configurations including half-cells and symmetrical cells (introduced in Section 4.2). Finally, a corresponding equivalent circuit is proposed and the identified loss mechanisms are assigned to electrochemical processes. The latter is then used to determine the parameter dependencies of each loss process.
The basic questions of this chapter are:

1. What is the advantage of DRT analysis for lithium-ion battery electrodes?
2. How can experimental lithium-ion cells be applied to identify electrochemical loss mechanisms reliably?
3. Which are the electrochemical loss processes dominating the investigated electrodes?

The main part of the results presented in this chapter have been published in Refs. [137, 144–146].

5.1 Electrodes and Measurement Setup

In this study $LiFePO_4$-cathodes fabricated at Fraunhofer ISC (Würzburg) were investigated. The cathodes were made of the basic active material $LiFePO_4$ (carbon coated) from Süd-Chemie, to which carbon black and poly (vinylidene fluoride)-binder (PVDF) were added to get a final weight ratio of 70:24:6 ($LiFePO_4$:carbon black:binder). After adding N-Methyl-2-pyrrolidon (NMP), the slurry was coated with a doctor blade on aluminum foil and vacuum dried for two hours at $80\,°C$. Then the electrode samples were stamped out and again dried for three hours at $120\,°C$.
The resulting layers shown in Figure 5.1 had a low thickness of around $20\,\mu m$. The microstructure was analyzed in Ref. [147] by FIB/SEM tomography and subsequent image processing in order to determine microstructure parameters. A very high porosity of $65.6\,\%$ and a high carbon black volume fraction of $19.4\,\%$ were measured. Only $15\,\%$ of the electrode volume contains the cathode active material $LiFePO_4$.

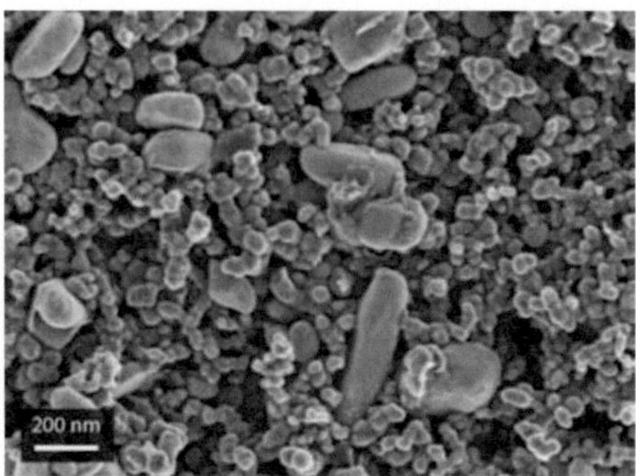

Figure 5.1: SEM-image of the lab-scale $LiFePO_4$-cathode investigated [147]. The green color highlights the $LiFePO_4$ particles and the small particles in between represent the carbon black added to increase the electronic conductivity.

These electrodes are very different compared to state-of-the-art electrodes used in today's lithium-ion batteries as they provide a very high carbon black content, a much lower layer thickness and a drastically higher porosity [147]. The last two characteristics are caused by not compressing the dried lab-scale electrodes. This additional step called calendering is usually applied in order to reduce the porosity of the electrode structure [148]. From a theoretical point of view, these microstructure properties indicate an excellent electrode performance in terms of high discharge rates because a low layer thickness, a high porosity and a high carbon black content minimize the ion and electron transport losses.

The anode material was lithium metal foil from Sigma Aldrich with a thickness of 0.38 mm. Both electrodes were separated by a 200 μm thick glass fiber separator (Freudenberg FS2019) serving as electrolyte reservoir. The liquid electrolyte was composed of a one molar $LiPF_6$-solution in a 1:1 mixing ratio with ethylene carbonate:ethylmethyl carbonate (EC:EMC). $LiFePO_4$-cathodes and lithium metal anodes were combined as half-cell configuration in EL-Cell housings [135] with an active electrode area of $2.54\,cm^2$. The cell assembly was carried out under Argon atmosphere in a glovebox (MBraun).

After assembling, ten charge and discharge cycles were performed with each half-cell to ensure a steady state of the system. The cycling obeyed a CCCV protocol using the theoretically calculated C/2 rate as charge and discharge current. The expected cathode capacity is calculated from electrode mass, mass fractions and theoretical capacity of $LiFePO_4$. Figure 5.2 illustrates ten formation cycles for a $LiFePO_4$ half-cell. The discharge capacity reaches a stable value around $255\,\mu Ah/cm^2$ which is 17 % below the theoretically expected capacity of $307\,\mu Ah/cm^2$, calculated from electrode mass, composition and the theoretical capacity of $170\,mAh/g$. The charge capacity reveals a considerable gap between the first charge and discharge cycles, indicating an incomplete lithium intercalation during the first discharge cycle compared to the synthesized $LiFePO_4$. It is known that for $LiFePO_4$ the practically usable capacity is approximately 5 % below the theoretically usable capacity [9]. This explains only a part of the lower capacity achieved. An incomplete exploitation of active material in the electrode structure can furthermore result from isolated or not accessible particles due to a bad connectivity of the electrode structure or pores which are not filled with electrolyte. In this study, usually between 80 and 90 % of the expected capacity could be reached.

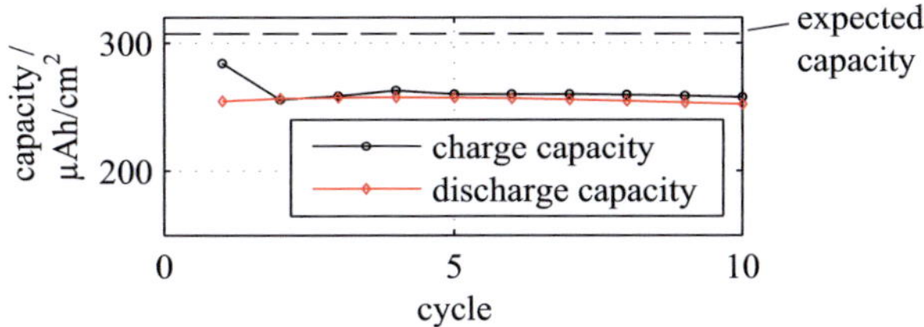

Figure 5.2: First cycles of a LiFePO$_4$ half-cell.

The cycled half-cells were afterwards investigated by EIS or opened in the glovebox and their electrodes re-assembled in symmetrical cells. Impedance measurements were carried out using the Solartron 1400E cell test system with Scribner Multistat software. The amplitude of the ac voltage applied was 10 mV RMS under open circuit condition and the frequency was varied within a range of 1 MHz to 10 mHz. Data consistency and quality was controlled by checking the Kramers Kronig residuals of all impedance spectra (presented in Section 4.5). After ten formation cycles, impedance spectra were measured for varying temperatures and SOCs in order to investigate the characteristic parameter dependencies. The SOC was stepwise decreased from 100 % to 10 %, whereas it was defined by the integration of discharge current and the previously measured discharge capacity. At each SOC, the temperature was varied between 0 °C and 30 °C using a Weiss WK1180 climate test chamber.

5.2 Impact of Temperature and SOC

5.2.1 LiFePO$_4$ Half-Cells

Figure 5.3 displays a series of impedance measurements of LiFePO$_4$ half-cells with SOC 100 % at temperatures of 30, 23, 20, 10 and 0 °C. The decrease of temperature gives rise to a consistent increase of total polarization as well as of ohmic resistance. The characteristic shape of the impedance curve is modified, indicating a different temperature dependence of the underlying loss mechanisms. An additional semicircle emerges between 1 Hz and 100 Hz for low temperatures. Furthermore, the gap between this semicircle and the low frequency branch increases. The slope of the low frequency branch remains constant but the absolute polarization increases for each frequency point.

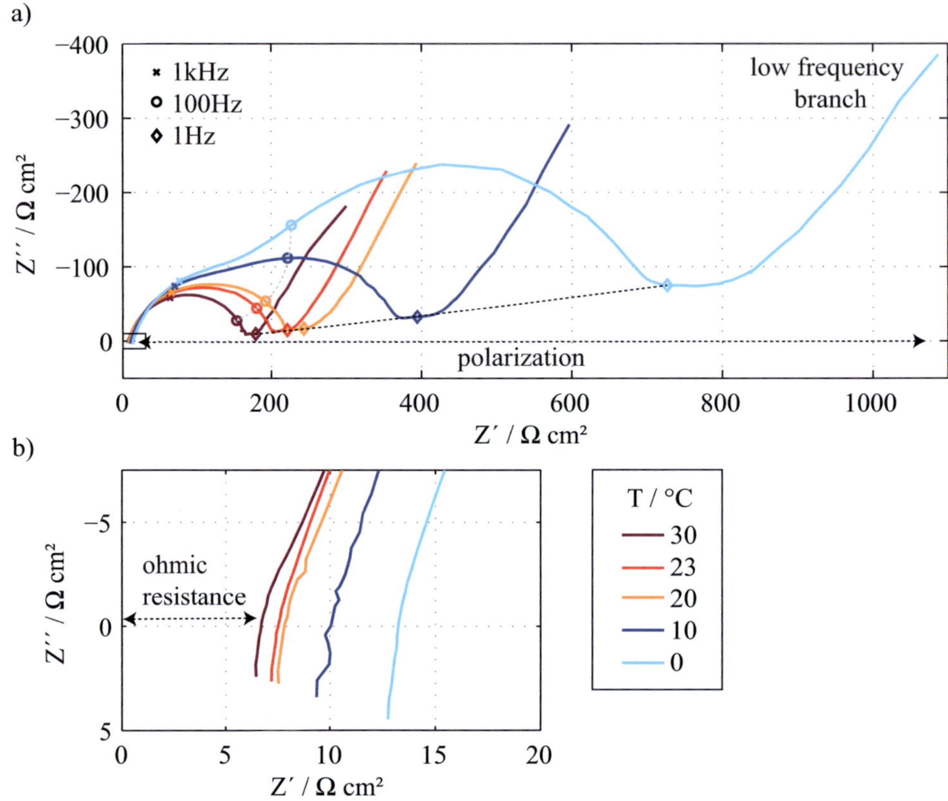

Figure 5.3: a) Nyquist plot of LiFePO$_4$ half-cell impedance spectra for varying temperatures at 100 % SOC. b) Enlargement of high frequency region.

The characteristic SOC-dependency of the half-cell impedance spectra is illustrated in Figure 5.4. The ohmic resistance reduces slightly for decreasing SOC although a constant ohmic resistance is expected. This reduction can be caused by an ongoing electrolyte distribution inside the pores and the separator for increasing measurement time. It stops to reduce for the last three impedance spectra (SOC=50,30,10 %) as they were measured at the end of the parameter variation. The overall polarization increases continuously for lower SOC due to a slight increase of the semicircle at high frequencies and a strong increase of the semicircle at frequencies below 100 Hz. Furthermore, the capacitive branch at low frequencies changes drastically. It becomes dominant for low SOC which can be explained by the SOC-dependent intercalation capacity $C_{\Delta int}$ (see Section 2.3.2).

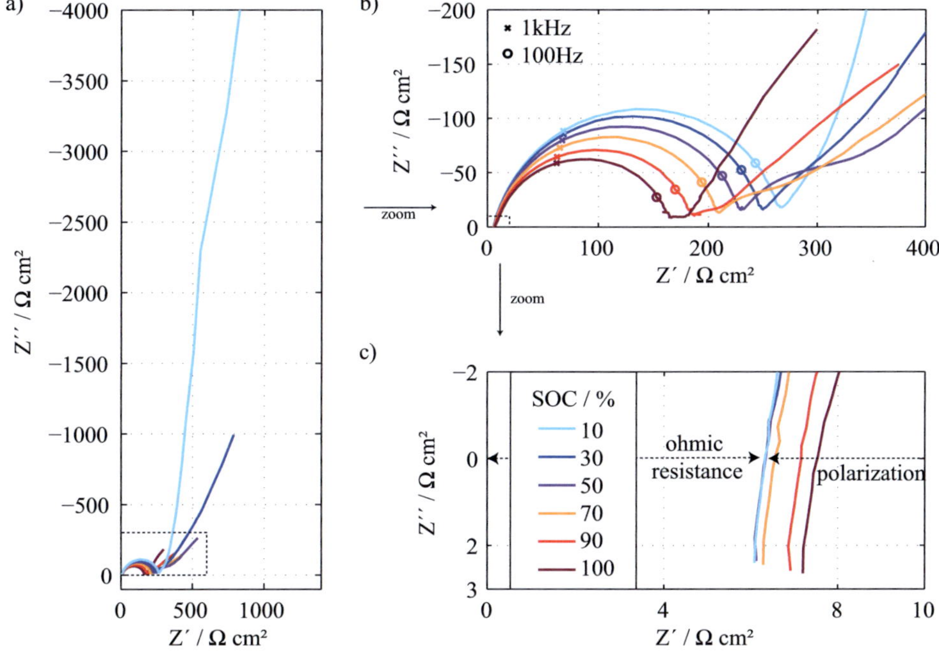

Figure 5.4: a) Nyquist representation of $LiFePO_4$ half-cell impedance spectra for varying SOC at 23 °C. Enlargement of b) medium and c) high frequency region.

5.2.2 Symmetrical LiFePO$_4$ Cells

Symmetrical LiFePO$_4$ cells were measured in order to separate anode and cathode losses. They were assembled from LiFePO$_4$-cathodes previously cycled and set to a certain SOC in half-cells. A change of the SOC is not possible after assembly as cycling changes both electrodes' SOC asymmetrically. Figure 5.5 depicts the Nyquist plot of a symmetrical LiFePO$_4$ cell for varying temperatures at 100 % SOC. The impedance spectra are divided by two in in order to represent the area-specific resistance for one LiFePO$_4$-electrode. This step results in a smaller ohmic resistance for the symmetrical cell compared to the half-cell (Figure 5.5c). Yet the temperature dependency of both cells is comparable. Furthermore, the symmetrical cell reveals two distinct semicircles of which the one at lower frequencies discloses a higher temperature dependency. The capacitive branch changes its polarization but not its slope, similar to what was observed for the half-cells' temperature variation.

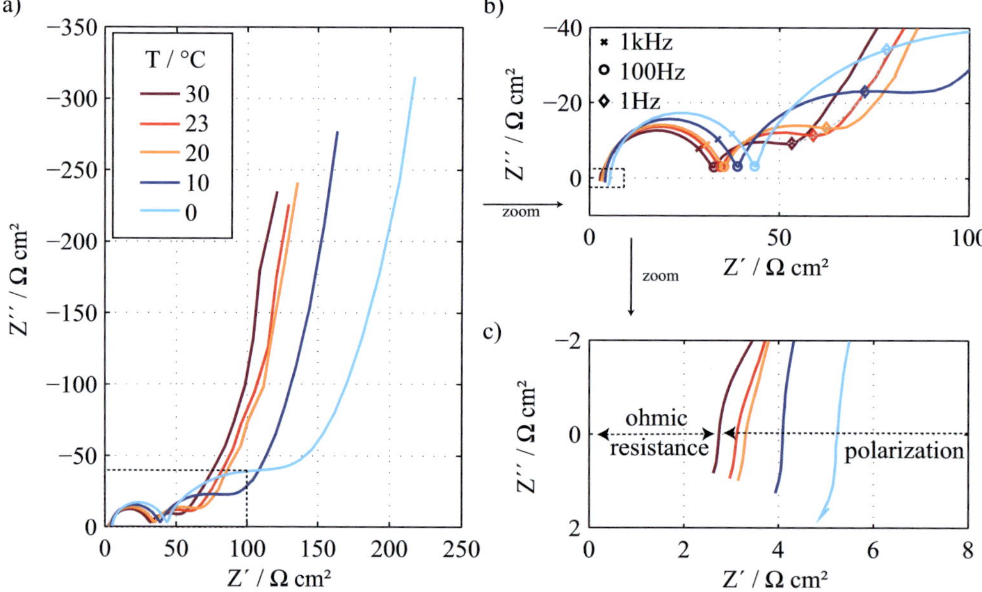

Figure 5.5: a) Nyquist representation of symmetrical LiFePO$_4$ cell impedance spectra for varying temperatures at 100 % SOC. Enlargement of b) medium and c) high frequency region. The spectra are divided by two in order to represent the area specific resistance of one LiFePO$_4$-electrode.

5.2.3 Symmetrical Lithium Cells

Symmetrical lithium metal cells were measured as well for a later assignment of half-cell loss processes. A Nyquist plot of such a cell for different temperatures is displayed in Figure 5.6. As already described above, lithium metal anodes were taken from cycled LiFePO$_4$ half-cells in order to get comparably prepared lithium metal anodes. The anodes presented here were cycled for three times in half-cells with a capacity of 250 µA and a current rate of C/2 and afterwards assembled as a symmetrical lithium cell. The corresponding impedance spectra comprise of one dominant and a second, very small and flat semicircle at lower frequencies, both showing a strong dependency on the temperature. The ohmic resistance is of comparable size and shows a similar temperature dependency as in the symmetrical LiFePO$_4$ cell.

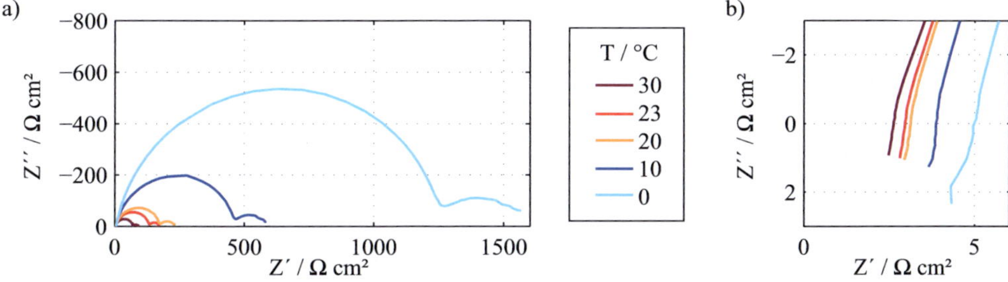

Figure 5.6: a) Nyquist plot of symmetrical lithium metal cell impedance spectra for varying temperatures. b) Enlargement of high frequency region. The spectra are divided by two in order to represent the area specific resistance of one lithium metal anode.

In Section 2.1.2.2 it was stated that lithium is deposited irregularly on the lithium metal anode surface during cycling. Therefore, cycling is expected to affect the impedance spectrum of experimental cells using lithium metal anodes. Some fundamental investigations have been conducted in this thesis in order to obtain a better understanding of the lithium metal anode impedance spectra.

First, a symmetrical cell was assembled from fresh lithium metal anodes without cycling the electrodes in advance. With that, impedance spectra were measured for the fresh cell and after several cycles in which the cell was cycled with 1 mA for one hour in both directions. After each cycle of charging the cell in both directions, an impedance spectrum was measured. The resulting spectra are presented in Figure 5.7. The characteristic shape differs slightly compared to the previously shown spectra as the low frequency part is more similar to a straight line than a semicircle. However, no fundamental difference can be observed. The cell shows the largest polarization for the first cycle, whereas the last spectrum reveals the minimal polarization. The decreasing polarization can be explained by an increasing surface area caused by the previously described irregular deposition of lithium during cycling. In summary, the polarization

of the impedance is reduced but its shape does not change fundamentally due to the
1 mA cycling. It is assigned to the lithium/electrolyte interface.

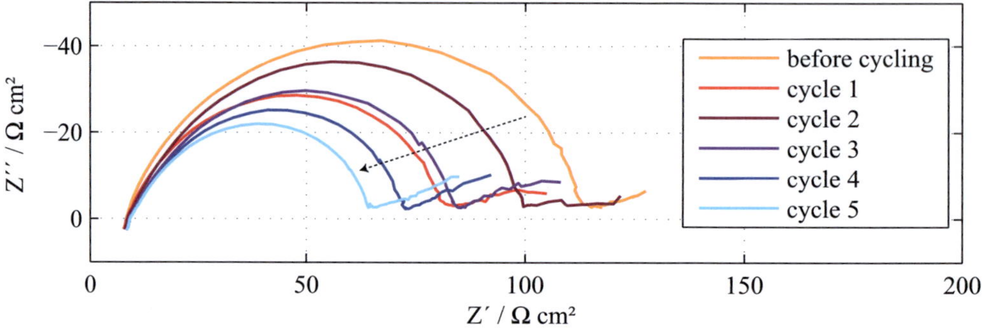

Figure 5.7: Nyquist plot of symmetrical lithium metal cell impedance spectra for cycling
the cell with 1 mA at T=23 °C.

Therefore, another symmetrical lithium cell was assembled from fresh lithium metal
anodes and afterwards cycled as described above. Thereupon, a variation of SOC
was simulated by charging the cell stepwise. Six charging steps were applied with
1 mA for one hour and all steps in the same direction. This would correspond to an
SOC-variation in half-cells with electrodes providing a capacity of 6 mAh. A copper
wire was put in the cell as a reference electrode and afterwards lithiated in order to track
the SOC-dependency of each lithium metal anode separately. Such a reference electrode
does not work absolutely reliably in terms a stable potential and the prevention of
artifacts [141], yet it allows for a qualitative visualization of SOC-dependencies. The
latter is illustrated in Figure 5.8 for the cell and both electrodes separately. The cell
impedance does not show a systematic change with SOC variation. It decreases steadily
due to the first four current steps. However, this trend turns and the impedance
increases strongly for the last two steps. It is more useful to have a look at the single
electrodes' impedance spectra because a more consistent trend can be observed.

The impedance spectrum of the first anode decreases continuously for each current
step. Anode two shows a slight increase after the first four current steps and increases
drastically for the last two steps. This behavior can be explained by the nature of
lithium deposition during cycling, illustrated in Figure 5.9.
In general, lithium accumulates at the surface of the lithium metal anode, forming a
porous lithium layer or several different porous layers at the anode surface [149]. The
latter increases the surface area available for lithium adsorption and dissolution. The
thickness of this porous layer depends on the amount of lithium transported during
cycling and therefore on the SOC of the symmetrical lithium cell.
The current pulses leading to the impedance spectra presented in Figure 5.8 transport
lithium from anode two to anode one as illustrated in Figure 5.9. Therefore, a porous
layer is formed on the first anode whereas it is decomposed on the second. Consequently,

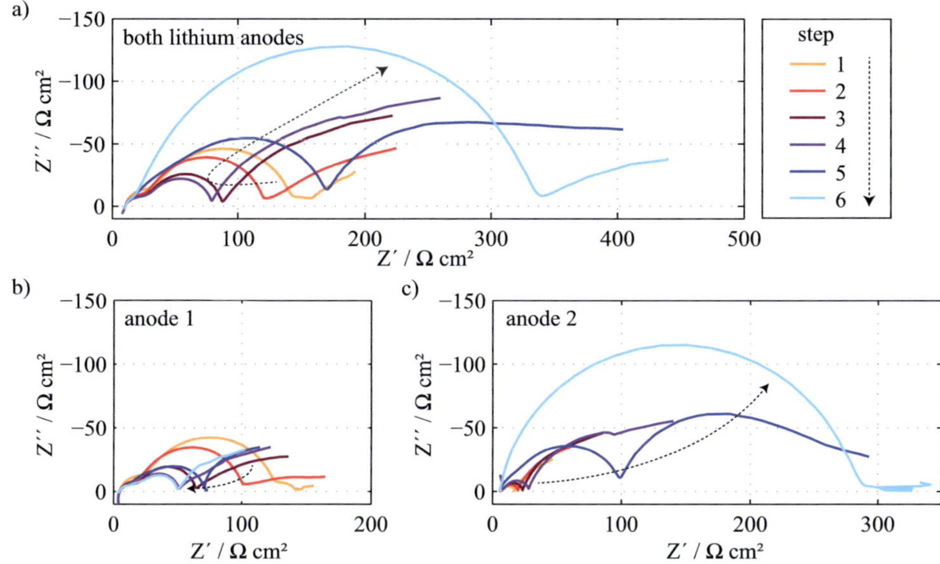

Figure 5.8: Nyquist plots of symmetrical lithium metal cell impedance spectra for step-wise charging the cell with 1 mA in one direction. Impedance spectra of a) full cell and b,c) single anodes measured by reference electrode.

the impedance spectrum of anode one decreases whereas that of the second anode rises drastically when the porous layer is completely decomposed and fresh lithium has to be dissolved.

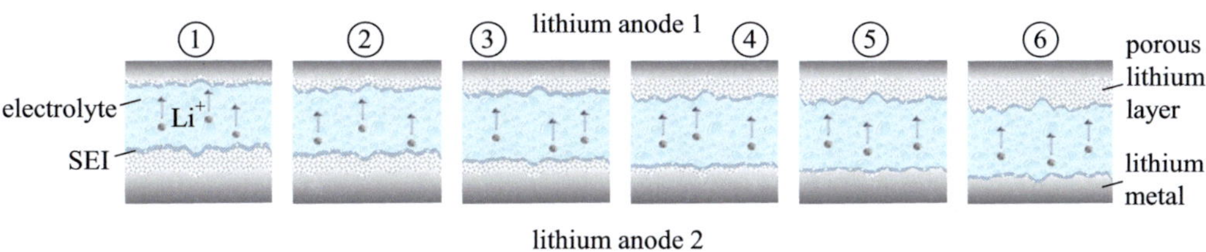

Figure 5.9: Scheme of lithium deposition during cycling in symmetrical lithium metal cells.

Not only the polarization resistance but also the shape of the measured impedance spectra changes drastically during these current step experiments. The dominant semicircle splits up into two or three semicircles, depending on the SOC of the cell (see Figure 5.8b and c). A large semicircle arises for low frequencies where only a small contribution was measured previously. This division into several loss mechanisms is related to the formation of the porous anode surface. It occurs when a large amount of lithium is cycled between the anodes and it can be related with a formation of more complex SEI layers and a resulting complication of the interface process.

In summary, the lithium metal anode has absolutely to be taken into account for the investigation of electrodes in half-cell configuration. Its polarization is directly related with the capacity of the investigated electrode and its cycling history. A consideration is more simple for low capacity electrodes (below 1 mAh per cell) as in this case, the lithium impedance spectra are more simple. However, it is more complicated to take the lithium metal anode into account during the analysis of electrodes with higher capacities. It changes its impedance spectra fundamentally and therefore prevents a reliable assignment and separation of loss processes. Generally, it is therefore preferable to apply a reference electrode in order to investigate both electrodes separately. This is why a reference electrode setup was developed later in this thesis in order obtain a reliable separation of losses even for high capacity electrodes.

5.3 Pre-Processing of Impedance Spectra

In the past, the DRT method was exclusively applied for the impedance model design and evaluation in SOFC research [70–72]. In this thesis it is used for the impedance model development for the first time in the field of lithium-ion batteries. The problems arising from this and ways to adapt the DRT method to be suitable for lithium-ion cells are introduced in this section. The fundamental difference between SOFC and lithium-ion cell impedance spectra is related to their low frequency behavior. The standard range for EIS usually covers frequencies down to 10 mHz in order to avoid extremely long measurement times accompanied by a violation of the stationary condition. These frequencies are not sufficient to stimulate all loss processes occurring in lithium-ion batteries. Especially the solid state diffusion is too slow to be fully recorded by these frequencies. Furthermore, lithium-ion batteries always show a capacitive behavior for low frequencies due to the intercalation capacity (see Section 2.3.2). These two facts lead to an increasing negative imaginary part for the low frequencies of lithium-ion cell impedance spectra, never dropping towards the real axis.

The effect of such a capacitive system on DRT-calculation is demonstrated in Figure 5.10. A DRT is calculated from the impedance spectrum presented, illustrated in Figure 5.10b. Next, this DRT is transformed backwards to its corresponding impedance spectrum and also displayed in the Nyquist plot in Figure 5.10a. It is obvious that the two impedance spectra do not concur. A better visualization for the difference between the latter two is provided by the DRT residuals, representing the relative deviation of real and imaginary part. These residuals are calculated just as the residuals between impedance measurements and impedance model, using Equation 2.22 and 2.23 in Section 2.2.4. The impedance represented by the DRT corresponds to the impedance spectrum which can be described by the calculated DRT. The residuals in Figure 5.10c reveal that the real part of the spectrum can be approximated very well. This originates from the DRT calculation which is based on the real part of the impedance spectrum. However, the imaginary part cannot be described exactly for high frequencies and is even less exact for low frequencies. This raises the issue of DRT interpretation in these frequency regions. It is known that the influence of cable inductance increases for high frequencies. This is not critical as a pure inductance does not influence the DRT result. However, it is not clear for low frequencies if the underlying electrochemical loss processes are represented well as the origin of the capacitive branch can be purely capacitive or it can include other low frequency loss mechanisms.

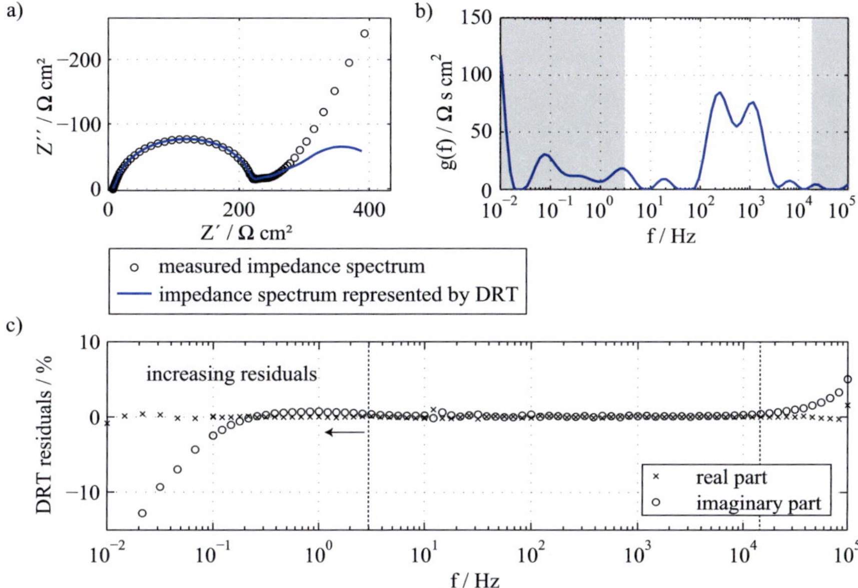

Figure 5.10: a) Nyquist plot and b) DRT of the measured LiFePO$_4$ half-cell impedance spectrum (black circles) at 20 °C and SOC=100 % and the spectrum which is represented by the DRT calculated from it (blue line). c) Real part and imaginary part of the residuals calculated from the deviation between the two impedance spectra.

Simulations have been conducted in order to elucidate the issue of DRT-interpretation for low frequencies. Therefore, a simple impedance model illustrated in Figure 5.11 was set up in serial connection. It consists of two RQ-elements, one 1D FLW and an intercalation capacity. Two frequency ranges were chosen for simulation, one of them from 1 MHz down to 10 mHz as usually investigated by EIS and another one down to 100 µHz.

The results of the first simulation are presented in Figure 5.12. The Nyquist plots of the entire model, the model without capacity and only the two RQ-elements are illustrated. The RQ-element's characteristic frequencies are of three decades difference and the Warburg element overlaps with RQ_2. A clear separation exists between high and low frequency processes. The evaluation of these impedance spectra by DRT in Figure 5.12c leads to three central conclusions. First, the RQ_1-peak at high frequencies is not

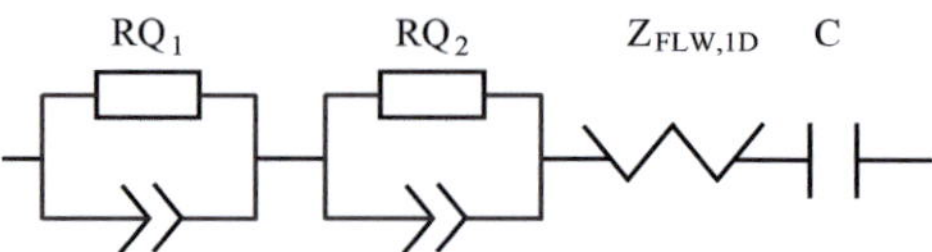

Figure 5.11: Equivalent circuit model used for the demonstration of DRT application in lithium-ion cells.

influenced by the other low frequency losses, leading to an unambiguous identification. Second, the identification of RQ_2 is impeded by solid state diffusion and capacity. The corresponding peak is inseparable as it is overlapped by the latter two. In the end, the DRT-calculation is not affected by the capacity. This is once more due to the DRT-calculation from the real part of the impedance spectrum.

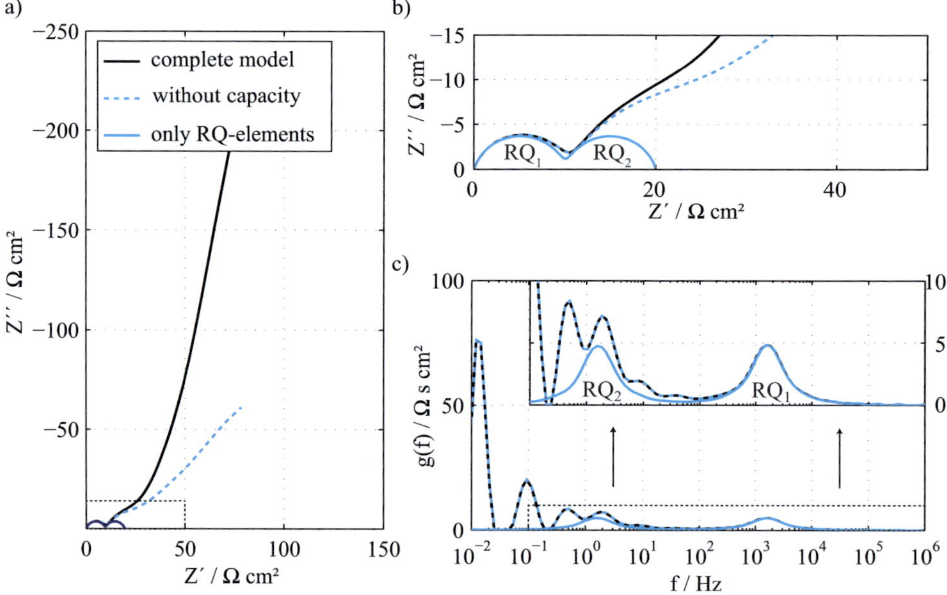

Figure 5.12: a)/b) Nyquist plot of the simulated impedance model shown in Figure 5.11. The complete model (two RQ-, one FLW element and a serial capacity), the model without capacity and only the two RQ-elements are simulated. c) DRT calculated from the simulated impedance spectra. The simulations cover frequencies between 1 MHz and 10 mHz.

The second simulation in Figure 5.13 was conducted in order to demonstrate the capability of measurements to very low frequencies. The extended measurement range does not affect the high frequency region compared to the previous simulation. Similarly,

RQ_1 is well-separated whereas RQ_2 overlaps with solid state diffusion and the capacity does not have an effect on the DRT. However, the identification of solid state diffusion is affected. Figure 5.13a discloses that a simulation without capacity allows for an identification of the typical solid state diffusion impedance. The imaginary part approaches zero for low frequencies, which indicates that the latter is recorded completely by the measurement. This allows for an evaluation of the corresponding DRT in Figure 5.13c which represents the true peak sequence of the solid state diffusion. The time constant of the underlying Warburg element can by identified reliably by the main peak in the DRT.

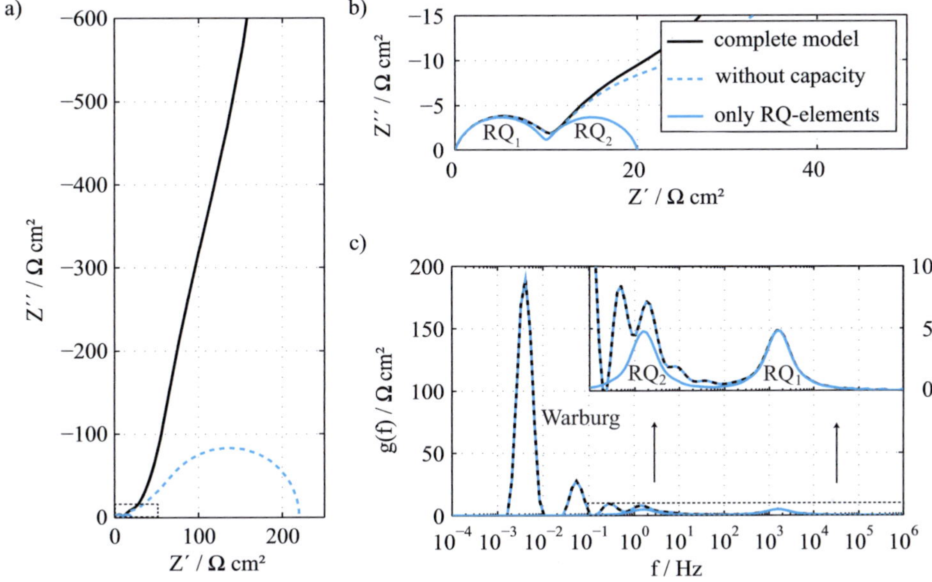

Figure 5.13: a)/b) Nyquist plot of the simulated impedance model shown in Figure 5.11. The complete model (Two RQ-, one finite length Warburg element and a serial capacity), the model without capacity and only the two RQ-elements are simulated. c) DRT calculated from the simulated impedance spectra. The simulations cover frequencies between 1 MHz and 100 µHz.

Three possible procedures follow from these findings for the application of the DRT in lithium-ion cells:

1. The DRT can be calculated for the entire impedance spectrum. It is then evaluated only for frequencies before the characteristic bend, neglecting the low frequency loss processes.

2. The impedance spectrum is pre-processed before the DRT-calculation by subtracting the solid state diffusion and the intercalation capacity. This allows for an

evaluation of the low frequency loss processes usually covered by the solid state diffusion.

3. The frequency range is extended and the intercalation capacity is subtracted subsequently. This allows for an unambiguous identification of solid state diffusion if the frequencies measured are sufficiently small.

The first procedure will be applied in Section 7, whereas Section 8 demonstrates the application of procedure three. In this section, the second method will be used. The latter is explained in the following in more detail.

The pre-processing requires a physically motivated low frequency model, which is then fitted to the low frequency branch and subtracted from the impedance data. As stated in literature, the low frequency branch originates from the solid state diffusion in the electrodes [59, 150]. Hence, in what follows we refer to it as *capacitive diffusion branch* and use two kinds of diffusion models conceivable for $LiFePO_4$-electrodes. The first model applied was introduced by Levi et al. [59]. It consists of a one-dimensional Generalized Finite Length Warburg element (1D GFLW) and a serial capacitor and models the diffusion adequately. Alternatively, the one-dimensional Generalized Finite Space Warburg element (1D GFSW) is applied in order to compare their suitability. The latter describes both, the solid state diffusion and the capacitive behavior, and does therefore not need a serial capacity. Both diffusion elements are selected in the one-dimensional form as this represents the one-dimensional diffusion in the $LiFePO_4$ adequately.

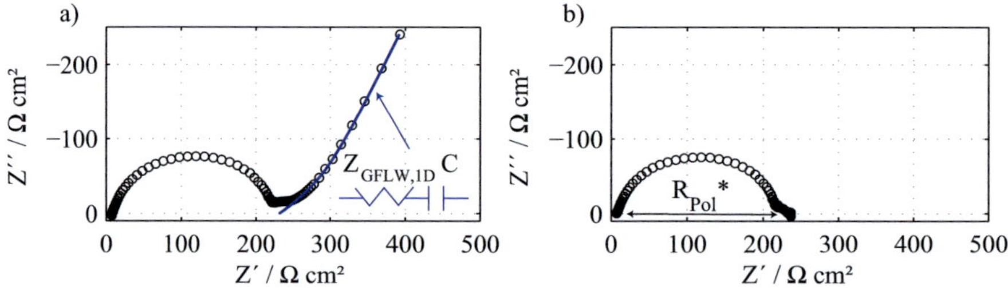

Figure 5.14: a) Nyquist plot of measured $LiFePO_4$ half-cell impedance spectrum (black circles) at 20 °C and SOC=100 % and impedance spectrum of the diffusion model approximated (blue line). b) Pre-processed impedance spectrum after fit and subtraction of the diffusion model.

The procedure which follows after choosing the correct diffusion element is demonstrated in Figure 5.14. The latter is approximated by a CNLS-fit to the low frequencies (in this case below 0.1 Hz) of the measured impedance spectrum, as a sufficient number of frequency points is necessary for the fitting procedure. Moreover, this procedure minimizes the influence of loss processes occurring at higher frequencies than solid state diffusion. These fits of the separate diffusion model are used for the pre-identification of loss processes and the model development. For the parameter dependency evaluation presented in the following chapters, the complete model was fitted to the entire frequency range in order to exclude an influence of the previous frequency selection on the fit results.

Equation 5.1 explains the pre-processing procedure for the first diffusion model proposed.

$$\underline{Z}_{pre}\left(\omega\right) = \underline{Z}\left(\omega\right) - \underline{Z}_{GFLW,1D}\left(\omega\right) - \underline{Z}_{C}\left(\omega\right) = R_0 + R_{pol}^* \cdot \sum_{n=1}^{N} \frac{g_n}{1 + j\omega\tau_n} \qquad (5.1)$$

The approximated diffusion impedance is subtracted from the measured impedance spectrum. Having been pre-processed this way, the resulting impedance $\underline{Z}_{pre}\left(\omega\right)$ fulfills the conditions for a correct DRT calculation. As mentioned above, two diffusion models have been compared. Figure 5.15 shows the pre-processed impedance spectra. A clearly better quality of pre-processing can be achieved by the FLW and the serial capacity compared to the FSW model. The latter causes loop-like artifacts in the pre-processed impedance spectrum due to an overestimated diffusion impedance.

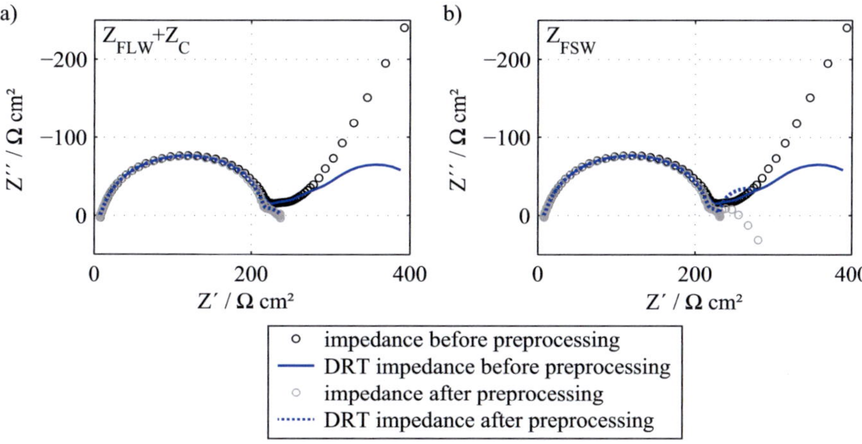

Figure 5.15: Nyquist plot of measured LiFePO$_4$ half-cell impedance spectrum at 20 °C and SOC=100 % (black circles), pre-processed impedance spectra (gray circles) and spectra which are represented by the DRT (blue line). a) Application of FLW and serial capacity and b) FSW as diffusion model.

This leads to large imaginary residuals for low frequencies, as presented in Figure 5.16. The pre-processing with FLW and serial capacity works much better, leading to very small residuals for real and imaginary components. Due to these better fitting results and a meaningful physical interpretation, the FLW and the serial capacity are used as diffusion model in the following sections.

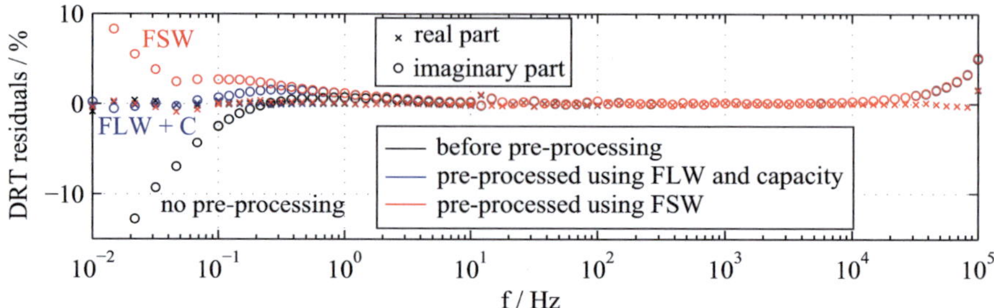

Figure 5.16: Real part and imaginary part of the residuals calculated from the deviation between the measured respectively the pre-processed impedance spectra (shown in Figure 5.15) and the spectra represented by the corresponding DRTs.

5.4 Identification of Loss Mechanisms by DRT

The measured impedance spectra of $LiFePO_4$ half-cells and symmetrical $LiFePO_4$ cells introduced in Sections 5.2.1 and 5.2.2 were pre-processed according to the procedure previously introduced. The symmetrical lithium cells were not pre-processed as they do not exhibit a capacitive diffusion branch. The spectra of all three cell configurations are presented in Figure 5.17 for a varying temperatures between 0 °C and 30 °C at SOC=100 %.

The low temperature impedance spectra of $LiFePO_4$ half-cell presented in Figure 5.17a reveal three characteristic semicircles. The series of pre-processed impedance spectra indicates that at least one of the three polarization processes is temperature-dependent. However, no information is available from this plot regarding polarization resistance and characteristic frequency of the underlying processes. Obviously, the DRT in Figure 5.18a gives considerably more detailed insight into the system than the pre-processed impedance plot. At least three polarization processes, P_1, P_2 and P_3, are clearly separable by the naked eye. Processes P_1 and P_2 are strongly temperature-dependent. They shift to lower frequencies and show increasing resistance for decreasing temperatures. In contrast, process P_3 shows no distinctive temperature dependency, but a rather large resistance. Besides a more sensitive separation of the underlying polarization losses, this clear visualization of temperature and frequency dependencies is an additional advantage of the DRT. If the DRT at 0 °C is studied in detail, a fourth process is discovered between P_2 and P_3, which overlaps with P_3 for higher temperatures. This process is neglected in this study because it is not separable from P_3 and plays a minor role compared to the other losses. This study focuses on the investigation of process P_1, P_2 and P_3 which represent the major loss processes. For a physical interpretation, an assignment to anode and cathode becomes necessary. Therefore, the pre-processed half-cell impedance spectra are compared to the symmetrical cells.

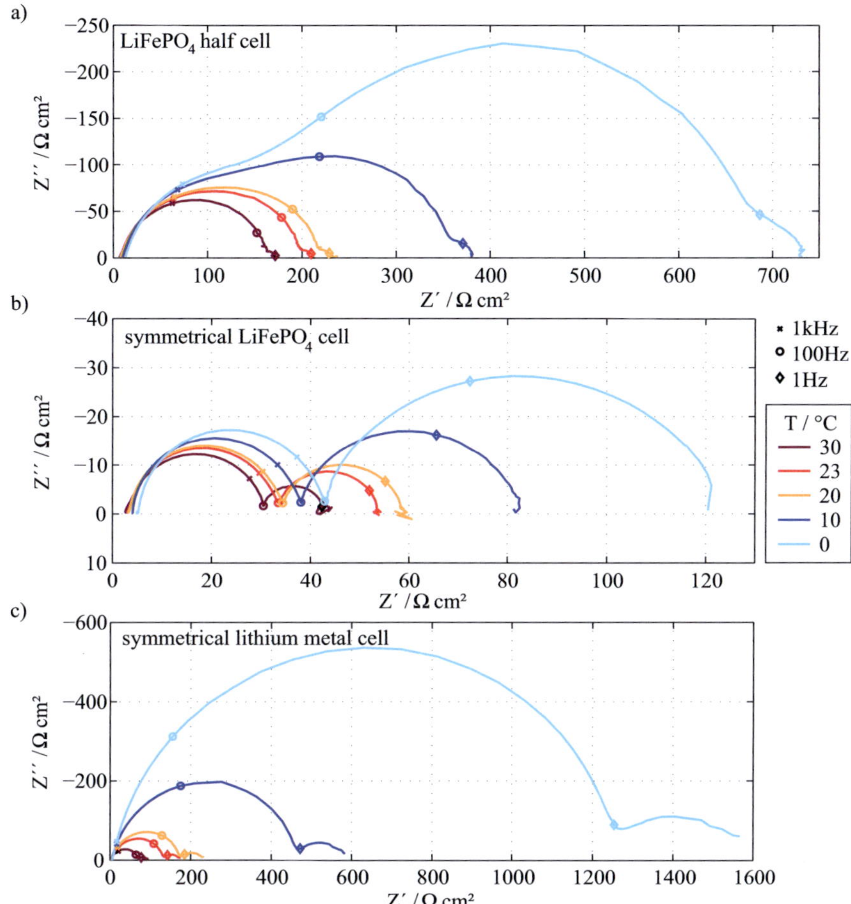

Figure 5.17: Nyquist plot of a) LiFePO$_4$ half-cell, b) symmetrical LiFePO$_4$ cell and c) symmetrical lithium metal cell for varying temperatures at 100 % SOC (for LiFePO$_4$-electrodes). The impedance spectra of LiFePO$_4$ half-cell and symmetrical LiFePO$_4$ cell were pre-processed as introduced above.

Figures 5.17 and 5.18 present all impedance curves and DRTs. In the case of symmetrical LiFePO$_4$ cells in Figure 5.18b, two processes are identified: Former P_1 occurs within a frequency range of 0.4 to 10 Hz and is temperature-dependent, and former P_3 remains around 3 kHz and is almost temperature-independent. The low frequency loss process P_1 shows one clear peak for high temperatures but it separates into two peaks for low temperatures. This behavior is possibly created by a small difference between both electrodes as the polarization processes can vary in frequency and scaling factor according to the inhomogeneities in electrode microstructure and lithium concentration. This effect makes the clear subtraction of the diffusion branch more difficult and leads to blurred DRT results. Moreover, P_1 and P_3 measured for the symmetrical cell differ slightly in their characteristic frequency and have a different polarization when compared with the $LiFePO_4$ half-cell. Yet they cover the same frequency range and show the same temperature dependency. Therefore, P_1 and P_3 are assigned to polarization processes caused by the $LiFePO_4$-cathode in the half-cell. Process P_2 in the medium frequency region is strongly temperature-dependent, as in the symmetrical lithium cell (Figure 5.18c), and is therefore assigned to the lithium-anode. In addition to showing P_2 as the major loss process, the DRT of symmetrical lithium cells shows some significantly less pronounced low frequency peaks. These peaks naturally also occur in the half-cell. However, their characteristic frequency range overlaps strongly with the solid state diffusion in the cathode, which makes separation impossible. This means that they are either subtracted with the capacitive diffusion branch or remain in the pre-processed spectrum. Hence, these clearly visible peaks influence either the obtained values for solid state diffusion or P_3. To this end, the symmetrical cell measurements and the corresponding DRTs enable to assign the $LiFePO_4$ half-cell processes P_1 and P_3 unambiguously to the cathode and P_2 to the anode. Therefore, as indicated in Figure 5.18b and Figure 5.18c, these loss processes are renamed P_{1A} (former P_2), P_{1C} (former P_1), and P_{2C} (former P_3) in the following sections.

Finally, it has to be discussed that the polarization processes P_{1A}, P_{1C}, and P_{2C}, when measured in varying cell configurations, differ slightly in their characteristic frequencies, their polarization and in temperature dependency. For the lithium loss process this is clarified by having a look at the section about lithium metal anodes (Section 5.2.3). These cells are highly variable due to their changing lithium electrolyte interface causing the differences between the symmetrical and the half-cell measurements. Furthermore, additional loss processes, as for instance between P_{1A} and P_{2C} in Figure 5.18a may be caused by the lithium metal anode. This is a critical point which has to be considered at the previously introduced evaluations. This point is overcome by a newly developed experimental cell configuration with reference electrode applied in Chapter 6. A difference occurs between symmetrical $LiFePO_4$ cell and $LiFePO_4$ half-cell as well. As noted earlier, their loss processes P_{1C} and P_{2C} differ in characteristic frequency and polarization. This is on the one hand an effect of re-assembling the $LiFePO_4$-electrodes from half-cells in symmetrical cells. The adhesion of the cathode layer on the current collector is a problem during re-assembly, leading to a changing interface between them. Furthermore, the charge transfer may be influenced by the evaporation of the electrolyte after opening the half-cell inside the glovebox. On the other hand, the used $LiFePO_4$-cathodes in half-cell and symmetrical cell were not from the same batch. Therefore, variations due to the manual fabrication process are possible. A systematic investigation of fabrication parameters will be introduced in Chapter 6. Nevertheless, it is already possible to assign the loss processes unambiguously to the electrodes due to their very characteristic parameter dependencies and the clear separation of loss processes by the DRT. Consequently, the benefit of the proposed route for impedance analysis, namely pre-processing followed by calculating the DRT, is conclusive.

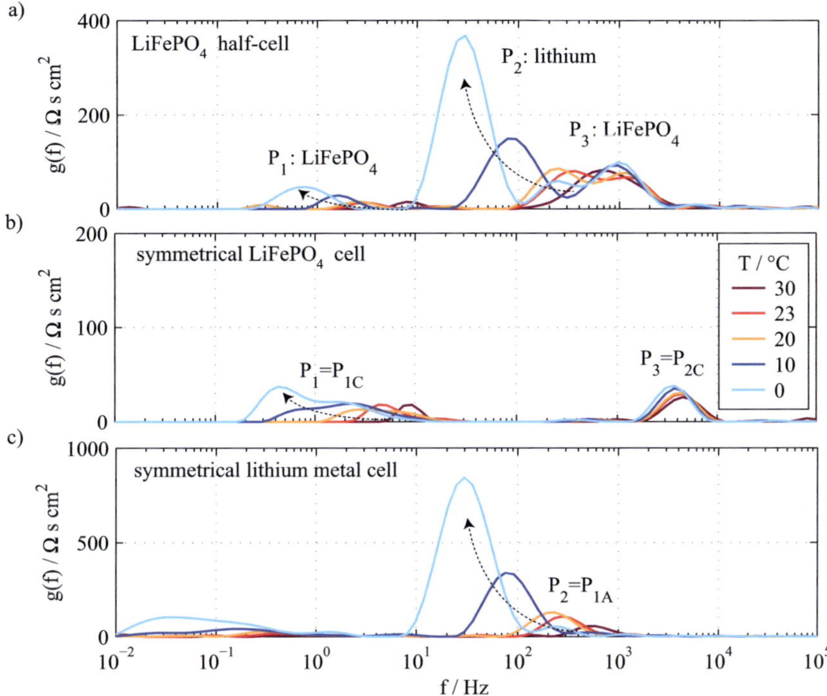

Figure 5.18: DRT of a) LiFePO$_4$ half-cell, b) symmetrical LiFePO$_4$ cell and c) symmetrical lithium cell for varying temperatures at 100 % SOC (for LiFePO$_4$-electrodes). The impedance spectra of LiFePO$_4$ half-cell and symmetrical LiFePO$_4$ cell were pre-processed as introduced above.

5.5 Equivalent Circuit Elements and Model Structure

The pre-identification of P_{1A}, P_{1C}, and P_{2C} in the previous section and the pre-processing of the capacitive diffusion branch generates the equivalent circuit model (ECM) in Figure 5.19. The proposed ECM consists of the one-dimensional generalized FLW element $Z_{GFLW,1D}$ with a serial capacity $C_{\Delta int}$, representing the solid state diffusion and the differential capacity of the cell. Furthermore, three RQ-elements are able to describe the polarization processes P_{1A}, P_{1C}, and P_{2C} as each of them represents one clear peak in the calculated DRTs. P_{1A} is assigned to the lithium metal anode, representing a combination of SEI-diffusion and charge transfer at the surface. The physical origin of the two cathode processes P_{1C}, and P_{2C}, is clarified in the following section. Lastly, the ECM contains a serial resistance R_0, summarizing the ohmic contributions of electronic and ionic charge transport phenomena in all cell components, and an inductance L, representing cable and cell inductance. The inductance is negligible for frequencies up to 100 kHz and therefore not relevant in the following discussion. The fitted impedance spectrum in Figure 5.19 shows a good approximation of the measured impedance spectrum.

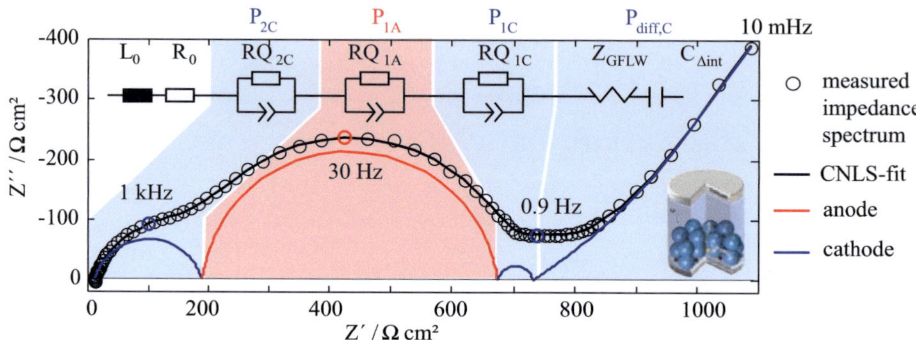

Figure 5.19: Nyquist plot of LiFePO$_4$ half-cell impedance for 0°C and SOC=100 %, obtained fit and equivalent circuit elements of the proposed impedance model (P_{2C}, P_{1C} and $P_{diff,C}$: cathode loss processes, P_{1A}: lithium loss process).

A better visualization of the fit quality is given by the DRT and the relative model residuals in Figure 5.20. The DRT demonstrates an excellent agreement between measurement and model for the entire frequency range, except the frequencies above 100 Hz. This is due to an imperfect modelling of the underlying lithium metal process P_{1A}. As extracted by the detailed investigation of symmetrical lithium cells (see Figure 5.18c), a comparable small, but additional loss process occurs at f >100 Hz. This process was already detected in LiFePO$_4$ half-cells for low temperatures between P_{1A} and P_{2C} (Figure 5.18a). The CNLS-fit becomes unstable concerning meaningful parameter dependencies when including this additional loss process into the model and is therefore neglected. Furthermore, this process overlaps with P_{2C}, caused by the LiFePO$_4$-cathode and cannot be separated for high temperatures. This affects the relative model residuals from the CNLS-fit with the proposed ECM, illustrated in Figure 5.20b. The residuals are below 0.5 % and do not show systematic deviations for all frequencies except the mentioned frequency range. Here, the residuals increase due to the additional peak not represented by the model, yet they still remain below 1 %. In the following section, the proposed ECM model is applied for analyzing the temperature and SOC dependence of the LiFePO$_4$ half-cell by CNLS-fit in Matlab.

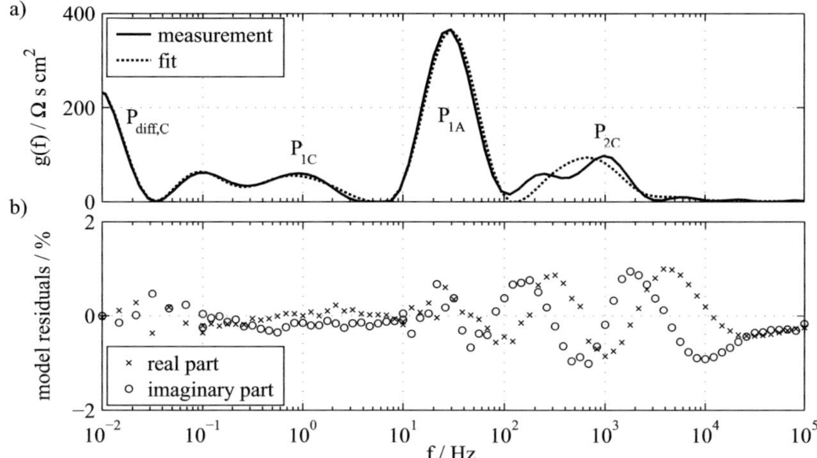

Figure 5.20: a) DRT of measured and fitted $LiFePO_4$ half-cell impedance for $0\,°C$ and SOC=100 %. b) Relative residuals for the corresponding model fit (P_{2C}, P_{1C} and $P_{diff,C}$: cathode loss processes, P_{1A}: lithium loss process).

5.6 Parameter Dependencies and Physical Interpretation

The CNLS-fit procedure profits from the DRT approach, as the DRT extracts the relaxation frequencies for each process, and thus delivers useful starting parameters. Figure 5.21 shows the area specific resistance (ASR) values and time constants of R_0, P_{1C}, P_{2C}, P_{1A}, and $P_{diff,C}$, as derived from the CNLS-fit, for the temperature range of $0\,°C$ to $30\,°C$ at SOC=100 %. The solid state diffusion $P_{diff,C}$ is dominant for all temperatures. It has a large time constant and reveals a distinct temperature dependency. P_{2C} is the second dominant loss process from cathode side. It takes place at the highest frequencies and is temperature-independent. The anode loss P_{1A} shows a strong temperature dependency and therefore exceeds the polarization of P_{2C} for temperatures below $10\,°C$. P_{1C} exhibits a very low polarization with a considerable temperature dependency. The ohmic resistance makes the smallest contribution to the impedance spectrum. A simple fit using Equation 2.7 delivers their individual activation energies, all showing Arrhenius behavior. Table 5.1 summarizes the characteristic frequencies F_R, the individual polarization values, denominated ASR, and the corresponding activation energy E_{act}. The ASR is normalized by the electrode surface area which is $2.54\,cm^2$. The ohmic resistance R_0 is connected with an activation energy of $0.14\,eV$, and in close agreement to $0.11\,eV$ for the conductivity of $LiFP_6$ in EC/DEC system and $0.155\,eV$ in EC/DMC. These values were reported in Ref. [151] for the measurement of temperature dependency of liquid electrolytes in polypropylene separators. Thus, we allocate the ohmic losses mainly to the ionic conductivity of the electrolyte system $LiFP_6$ in EC/EMC.

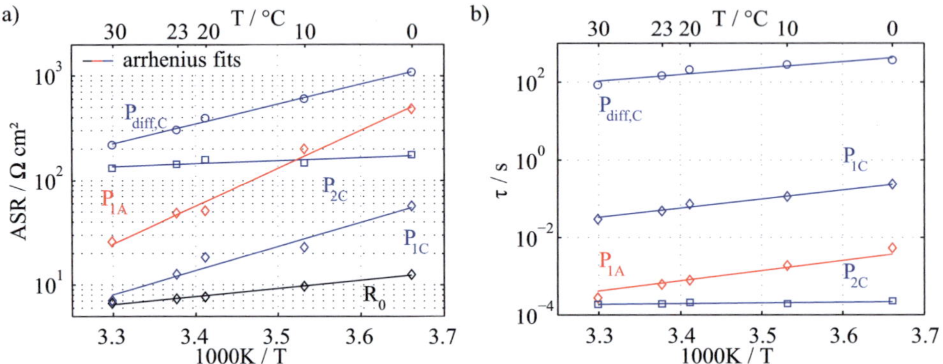

Figure 5.21: a) ASR and b) characteristic time constant of all identified loss mechanisms in a LiFePO$_4$ half-cell for varying temperatures at 100 % SOC (P_{2C}, P_{1C} and $P_{diff,C}$: cathode loss processes, P_{1A}: lithium loss process).

The anodic polarization P_{1A} shows an activation energy of 0.72 eV and is presumably linked to an SEI-diffusion process. In Ref. [152] activation energies varying from 0.3 eV to 0.83 eV were determined for the SEI-diffusion in different electrolytes. Thereby, the SEI was assumed to consist of several layers with separate activation energies. The value of 0.72 eV represents a cumulative temperature dependency of the complete SEI layer. The activation energy of the solid state diffusion, $P_{diff,C}$, is calculated from the diffusion coefficients at different temperatures which are not calculated here. The activation energy of P_{1C} is 0.45 eV and the activation energy of P_{2C}, being the major loss process for high temperatures, is 0.06 eV.

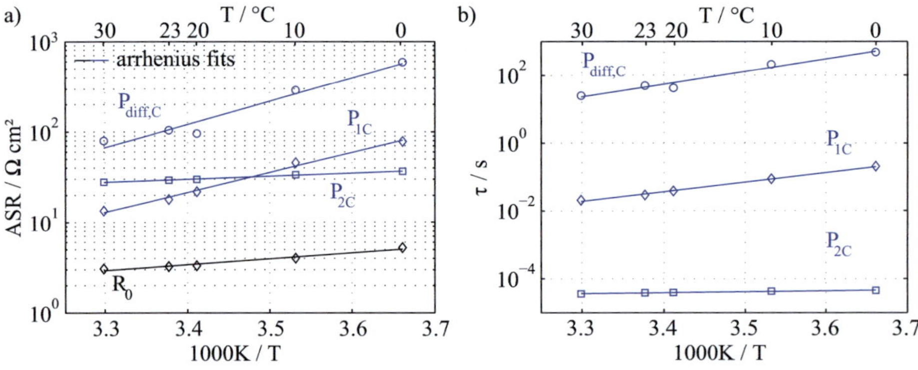

Figure 5.22: a) ASR and b) characteristic time constant of all identified loss mechanisms in a symmetrical LiFePO$_4$ cell for varying temperatures at 100 % SOC (P_{2C}, P_{1C} and $P_{diff,C}$: cathode loss processes, P_{1A}: lithium loss process).

process	equivalent circuit	LiFePO$_4$ half-cell			symmetrical LiFePO$_4$ cell		
		F_R/Hz	ASR/ $\Omega\,\text{cm}^2$	E_{act}/ eV	F_R/Hz	ASR/ $\Omega\,\text{cm}^2$	E_{act}/ eV
R_0	ohmic resistance	-	7 - 12	0.14	-	3 - 5	0.13
P_{1A}	RQ	500 - 30	26 - 486	0.72	-	-	-
P_{1C}	RQ	8 - 0.8	7 - 57	0.45	8 - 0.8	13 - 78	0.44
P_{2C}	RQ	1000	131 - 178	0.06	4.4 - 3.5k	28 - 37	0.07
$P_{diff,c}$	GFLW+C	2 - 0.4m	219 - 1089	-	6 - 0.3m	80 - 588	-

Table 5.1: Overview of identified loss mechanisms, applied equivalent circuit elements, temperature-dependent polarization and time constant from 30 °C to 0 °C and activation energies for LiFePO$_4$ half and symmetrical cells (P_{2C}, P_{1C} and $P_{diff,C}$: cathode loss processes, P_{1A}: lithium loss process).

The activation energies obtained from half-cell measurements were further cross-checked by symmetrical cell measurements in order to exclude a possible influence of the lithium metal anode on the evaluation. The temperature-dependent polarization and time constants of the symmetrical LiFePO$_4$ cells are presented in Figure 5.22. The relation between the polarization of P_{1C} and P_{2C} is different due to the reasons explained above. However, the temperature dependencies show the same behavior compared to the half-cell measurements. The resulting values for polarization, characteristic frequencies and activation energies are listed in Table 5.1 as well. The activation energies obtained from the symmetrical cell configuration are in excellent agreement with the values from half-cell measurements, thus proving the validity of the half-cell impedance model. According to the literature, charge transfer resistance is, apart from solid state diffusion, assumed to be the dominant loss process in LiFePO$_4$-cathodes [28, 108–110, 153]. Hence, either P_{1C} or P_{2C} is attributed to the charge transfer of lithium-ions at the interface cathode/electrolyte. However, as a charge transfer process is naturally temperature activated, this clearly points at P_{1C}, whereas the dominant P_{2C} originates from another cathodic process. It is not possible to address this question in the literature data as, the impedance response of the LiFePO$_4$-cathode was never before separated into several losses. Therefore, the investigation of P_{1C} and P_{2C} was continued by an SOC variation between 100 % and 10 %.

SOC-dependent polarization resistances and time constants are presented in Figure 5.23 for T=23 °C. The contributions show fundamentally different dependencies which can be separated very well by the impedance model evaluation. The ohmic resistance is almost constant as it was expected for the ionic resistance in the electrolyte. The lithium metal anode polarization resistance increases steadily with decreasing SOC. This is in agreement with the investigations of the lithium metal anodes in Section 5.2.3. During discharge, lithium is transported from the anode to the cathode side, reducing the surface of the lithium metal anode. This naturally increases the respective polarization. This hypothesis is supported by the SOC independent time constant for

P_{1A} as with this, the resulting double layer capacity $C_{DL} = \frac{\tau}{R}$ declines and indicates a reduction of the surface area as well. P_{1C} varies with SOC, backing up the fact that charge transfer is facilitated when few lithium-ions are present in the host lattice – hence, for high SOC. This is a supplementary indication that P_{1C} is caused by a charge transfer of lithium-ions across the interface cathode/electrolyte. In contrast, P_{2C} is totally constant for SOC variation, whereas the solid state diffusion $P_{diff,C}$ reveals a slight SOC dependency showing no systematic behavior. These trends also hold true for the corresponding time constants.

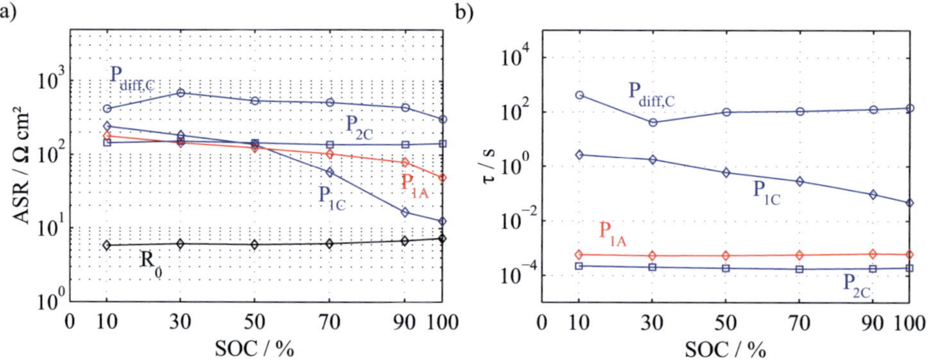

Figure 5.23: a) ASR and b) characteristic time constant of all identified loss mechanisms in a LiFePO$_4$ half-cell for varying SOC at T=23 °C (P_{2C}, P_{1C} and $P_{diff,C}$: cathode loss processes, P_{1A}: lithium loss process).

On the whole, the LiFePO$_4$ half-cell is analyzed at varying temperatures of 0 °C to 30 °C and SOC from 10 % to 100 %. The resulting field of operating conditions is then evaluated using the proposed equivalent circuit model and a CNLS-fit in Matlab. Now, the parameter dependencies of all cathodic processes, namely of charge transfer P_{1C}, a second polarization resistance P_{2C} and solid state diffusion $P_{diff,C}$ are clearly visible in Figure 5.24. As already presented in the previous paragraphs, the magnitude of P_{2C} is almost independent of temperature and SOC. The rather constant value of ASR=178 Ω cm^2± 50 Ω cm^2 strongly supports this polarization mechanism as being caused by the contact resistance between cathode and current collector (Figure 5.25) as Gaberscek et al. proposed in Ref. [117]. In contrast, the temperature- and SOC-dependency of the charge transfer cathode/electrolyte, P_{1C}, is pronounced. Its resistance varies by more than two orders of magnitude from ASR = 7 Ω cm^2 for 100 % SOC and 30 °C to ASR=1900 Ω cm^2 for 10 % SOC and 0 °C. This course is a further piece of evidence that P_{1C} is correctly attributed as the charge transfer process at the cathode/electrolyte interface. Lastly, the polarization of lithium diffusion in LiFePO$_4$ is shown in Figure 5.24b. As expected, $P_{diff,C}$ is very sensitive to temperature with ASR$_{min}$= 287 Ω cm^2 and ASR$_{max}$=1287 Ω cm^2 at an SOC of 50 %, but less sensitive to state of charge, with ASR$_{min}$=393 Ω cm^2 and ASR$_{max}$=909 Ω cm^2 at a temperature of 20 °C.

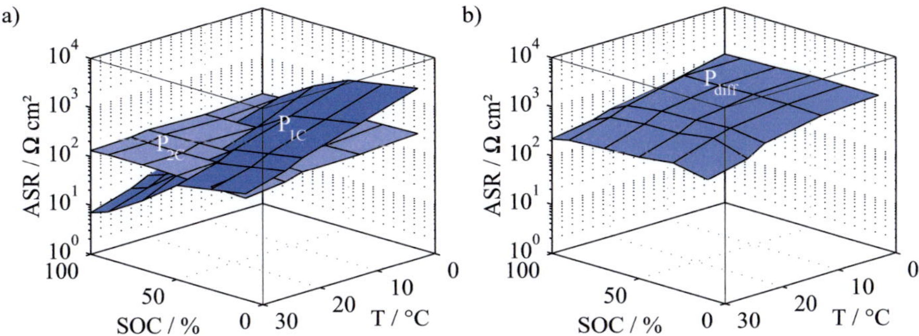

Figure 5.24: ASR depending on operating condition for a) Charge transfer P_{1C} between cathode and electrolyte, contact resistance P_{2C} between cathode and current collector and b) solid state diffusion $P_{diff,C}$ in the cathode active material.

Table 5.2 displays the values for P_{1C}, P_{2C} and $P_{diff,C}$, depending on operating conditions. The solid state diffusion, $P_{diff,C}$, is the major loss process in almost any case, followed by the contact resistance, P_{2C}, at high state of charge. The charge transfer resistance, P_{1C}, increases significantly at low values for temperature and SOC, representing the major part of the cathodic impedance for SOC=10 % and T=0 °C. The physical interpretation of all identified loss mechanisms is illustrated in Figure 5.25. A microstructure variation presented in Chapter 6 proves the physical interpretation of P_{1C} and P_{2C} and gives direction for further improvement of the cathode performance.

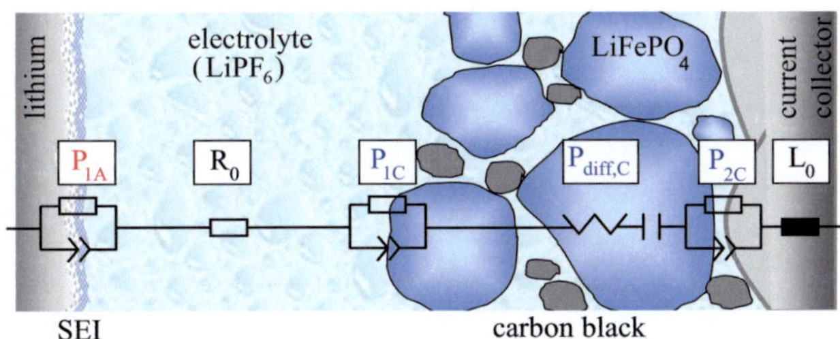

Figure 5.25: Illustration of all identified loss mechanisms and corresponding physical interpretation in LiFePO$_4$ half-cells including: The electrolyte resistance R_0, the contact resistance P_{2C} between cathode and current collector, the interface P_{1A} between lithium metal anode and electrolyte, the charge transfer P_{1C} between cathode and electrolyte, and the solid state diffusion $P_{diff,C}$ in the cathode active material.

Process	ASR / $\Omega\,cm^2$			Physical origin
	30 °C, SOC=100 %	20 °C, SOC=50 %	0 °C, SOC=10 %	
P_{1C}	7	196	1900	Charge transfer cathode/electrolyte
P_{2C}	131	180	225	Contact resistance cathode/current collector
$P_{diff,C}$	219	781	1308	Solid state diffusion

Table 5.2: Overview of identified cathode loss mechanisms, the corresponding polarizations and time constants for selected operating conditions and the physical interpretation.

5.7 Discussion and Conclusions

In this chapter, impedance measurements were conducted on $LiFePO_4$ half-cells in a wide range of operating conditions. A new method was introduced to apply the DRT for the evaluation of lithium-ion cells' impedance spectra by pre-processing the spectra before the application of the DRT method. The pre-processing step comprises the *a priori* selection of a low frequency impedance model, fitting the latter to the measured impedance and subtracting its resulting impedance spectrum from the measurement. Using this pre-processing and measuring symmetrical cells allowed for a clear separation of anode and cathode losses. The identification of the dominating loss mechanisms enabled the design of a physically motivated impedance model describing $LiFePO_4$-cathodes and lithium metal anodes in half-cell configuration. Finally, a physical interpretation could be achieved for all identified loss processes by determining their parameter dependencies using a CNLS-fit of the impedance model developed. With that it was possible to determine the parameter dependent polarization of each contribution. The identified loss processes for the $LiFePO_4$-cathode are: (i) solid state diffusion in the cathode, (ii) charge transfer resistance between cathode and electrolyte, and (iii) contact resistance between cathode and current collector. Depending on the parameters such as temperature and SOC, the order of magnitude of these dominant loss mechanisms changes.

The identification and interpretation of three contributions to the impedance of $LiFePO_4$-cathodes is unique compared to other studies. Most published papers assume only the charge transfer process to be relevant beside the solid state diffusion [28, 108–110, 153], although its dependency on cathode composition indicates that it could be the contact resistance as well [28]. There are some recently published papers identifying the contact resistance but not the charge transfer as loss process for $LiFePO_4$-cathodes [117, 154]. Only a few parameter variations are introduced to determine the corresponding dependencies of loss mechanisms on temperature [108] or SOC [110]. An entire field of both parameter dependencies is not available. A very new paper published by van

Bommel et al. [155] uses the results from this thesis already published in [137, 144–146] in order to investigate the performance of LiFePO$_4$-cathodes. As a result of Ref. [155], the interpretation presented in this thesis is confirmed and applied for the improvement of LiFePO$_4$-cathodes.

The cell configurations applied in this chapter were half-cells and symmetrical cells for the separation and characterization of cathode loss mechanisms. Thereby it was found out that the application of lithium metal anodes is limited concerning reproducibility due to its variable surface structure. The latter showed a dependency on SOC and also on cycling which is already known from Refs. [149, 156, 157]. This is caused by a continuous decomposition and new formation of SEI layer due to lithium deposition and dissolution. According to the literature, this process causes a complex SEI consisting of several layers which can be described well by at least four of RC-elements [149, 152, 158]. In this study, a simplified equivalent circuit model containing one RQ-element was applied in order to describe the lithium metal anode impedance. As presented in Section 5.2.3, this simplification is only valid under limited conditions and is error-prone to additional effects of the lithium metal anode. This error source has to be considered, which is the reason that symmetrical cathode cells were used in this study to cross-check the half-cell results.

In Chapter 6, a new cell configuration using a reference electrode is applied in order to exclude possible influences of lithium metal anodes.

6 Optimization of Lab-Scale Cathodes

A physically motivated impedance model for lab-scale $LiFePO_4$-cathodes was introduced
in the previous chapter. It allows for a quantification of all electrochemical processes
occurring in the cathode layer. In this chapter, the performance of these cathodes
under load is analyzed and compared to that of cathodes extracted from commercially
available cells introduced later in Chapter 8. Discharge experiments were conducted
with varying discharge current and evaluated by comparing the corresponding Ragone
plots (introduced in Section 2.1.5.2). Figure 6.1 represents the Ragone plots of a
lab-scale $LiFePO_4$-cathode as introduced in the previous chapter and a commercial
$LiFePO_4$-cathode. Energy and power density can be calculated in different ways, which
results in several diagrams.

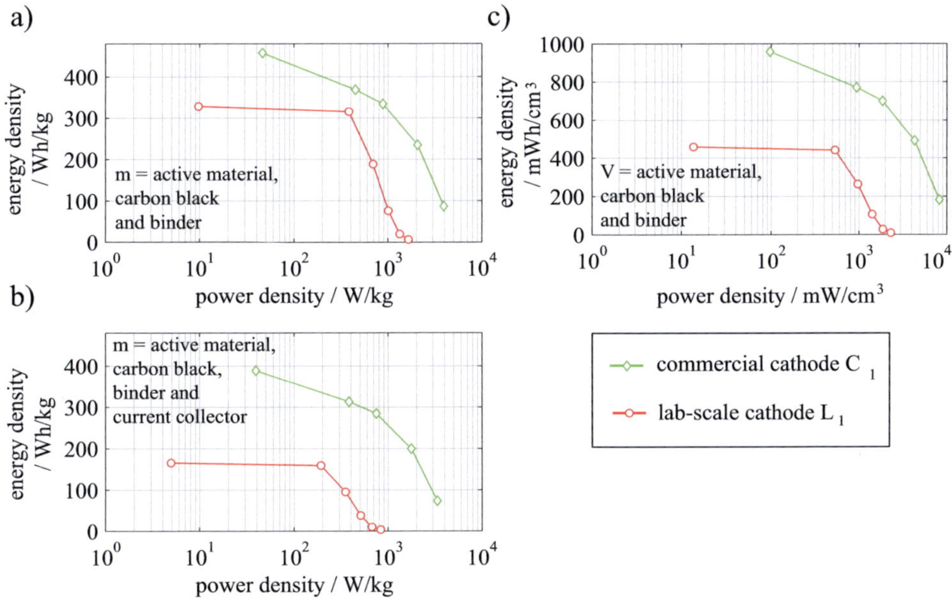

Figure 6.1: Ragone diagrams for lab-scale cathode L_1 and commercial cathode C_1
illustrating energy versus power density based on a) pure composite layer mass, b)
composite layer and current collector mass and c) composite layer volume.

Figure 6.1a depicts the Ragone plot in which the supplied energy and the average power
during a discharge cycle are referred to the pure mass of the cathode layer, comprising

the active material, carbon black and binder. The cathode layer is also denominated as composite layer due to its multiple components. Figure 6.1a reveals that energy as well as power density provided by the lab-scale are clearly lower than for the electrode from commercial cells. This becomes even more significant if the mass of the current collector is included in the calculation as is shown in Figure 6.1b. This larger difference is due to the lower area-specific loading of the here investigated lab-scale electrodes, leading to a higher impact of the additional inactive mass. Yet it is necessary to include the inactive components needed in a cell in order to obtain a fair comparison. Figure 6.1c illustrates the difference between the two electrodes for the volume-specific energy and power. Similarly, the commercial electrode provides far superior properties as its porosity and hence its volume is much lower which will be presented in the following section.

Therefore, the following chapter addresses the optimization of lab-scale electrodes in order to achieve a higher performance. For that, different electrode structures were fabricated at Fraunhofer ISC and analyzed in this study by impedance spectroscopy and the previously developed impedance model. A determination of their loss processes and their performance under load depending on the microstructure can be achieved. The major questions addressed in this chapter are:

1. What is the impact of electrode microstructure and fabrication parameters on the identified electrochemical loss mechanisms?
2. How do these mechanisms influence the performance under load?

6.1 Electrode Structures and Measurement Setup

Several parameters can be modified during the fabrication of composite electrode structures. An overview of the basic parameters is given in Figure 6.2. First, the layer thickness of the electrode coating can be adapted. A low layer thickness is expected to provide a good performance at high currents as electronic and ionic transport paths through the electrode layer are very short. Secondly, the size of active material particles can be varied, whereas small particles should allow for large currents as the lithium diffusion path inside the active material is then minimized and the active surface area for intercalation is increased. Then, it is possible to adjust the mass fractions of active material and conducting additives in order to tune the electronic conductivity of the composite layer. A high fraction of conducting additives generates a good electronic conductivity inside the electrode but it reduces the presumed energy density. Lastly, the adjustment of the porosity is presented in Figure 6.2 as a parameter. A high porosity supports an easy ion transport inside the pores but impedes the electronic conduction through the composite layer. The optimal value for the porosity is dependent on the other layer parameters and active material properties.

More complex electrode layers are fabricated for high performance cells in order to obtain superior electrode properties. For instance, active material agglomerates are introduced as a way to increase the energy density of electrodes without reducing their high current performance. Thereby, the active material particles agglomerate to larger units providing a low porosity, whereas a more generous porosity is available between these agglomerates for a fast lithium-ion transport. Additionally, adding carbon fibers (for instance VGCF) to the electrode produces a better percolation of electronic paths. The latter can also be achieved by increasing the networking of the carbon black particles during the preparation. Lastly, multilayer and gradient electrodes or the use of lithium-ion conducting binders are future methods for optimization.

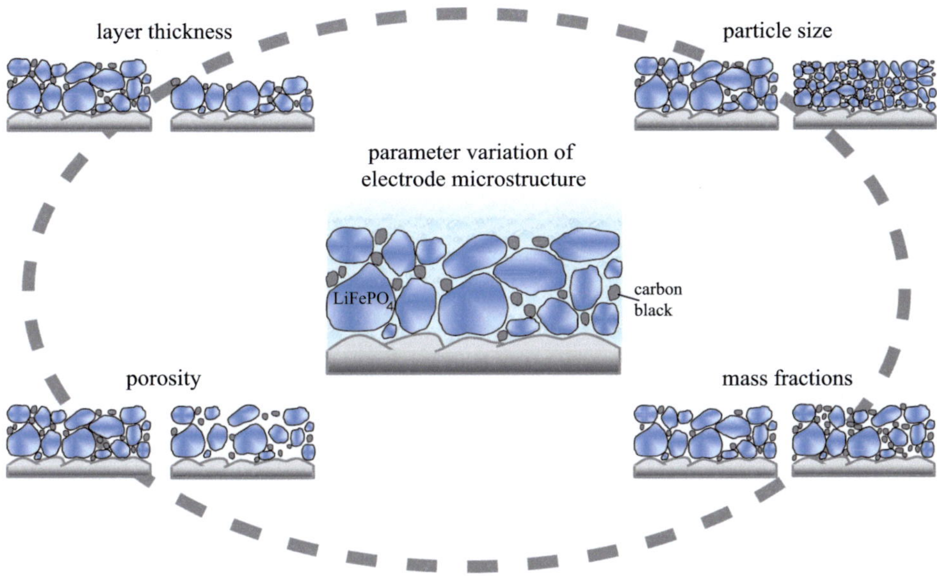

Figure 6.2: Basic parameters which can be adjusted in order to optimize a composite electrode for a certain application: Layer thickness, particle size, porosity and mass fractions (active material : carbon black : binder).

Different LiFePO$_4$ composite electrodes were fabricated at the Fraunhofer ISC (Würzburg) under systematic parameter variation in order to analyze the fundamental correlation between electrode microstructure and electrochemical performance. The electrode preparation followed the route described in the previous chapter. Four kinds of lab-scale cathodes are analyzed in this thesis. First, a cathode with the mass fractions of 70:24:6 (LiFePO$_4$:carbon black:binder) providing a very low layer thickness of 12 µm without calendering was chosen. Second, a calendered cathode layer with the same composition and layer thickness was applied in this comparison. Next, the mass fractions were adapted to a composition of 86:10:4 (LiFePO$_4$:carbon black:binder) for the same thickness in order to achieve more common cathode structures. Finally, the thickness was doubled using the same material composition. All these lab-scale electrodes were compared to a commercial cathode with a much larger thickness of 55 µm. The macroscopic properties

cathode ID	calendered	thickness / μm	mass fractions $(\mathrm{LiFePO_4:CB:PVDF})$
C_1	x	55	-
L_1	-	12	70:24:6
L_2	x	11	70:24:6
L_3	x	12	86:10:4
L_4	x	24	86:10:4

Table 6.1: Overview of the fabrication parameters of the cathode structures investigated.

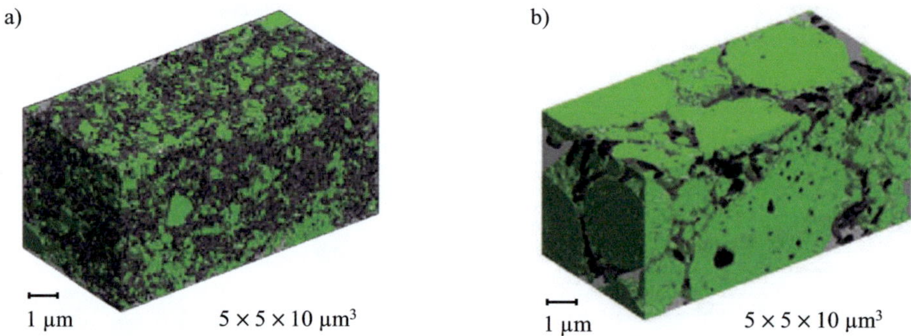

Figure 6.3: Representative volume element (RVE) of a) lab-scale cathode L_1 and b) commercial cathode C_1. The active material $\mathrm{LiFePO_4}$ is displayed in green, the small black particles represent the carbon black and the pore volume which is filled with electrolyte in the cell is transparent [147].

of these cathodes and the corresponding IDs used for the following analysis are listed in Table 6.1.

Having been fabricated as previously described, the microstructure of these electrodes was analyzed by Moses Ender at IWE by FIB/SEM analysis [147,159]. This enables the differentiation of $\mathrm{LiFePO_4}$, carbon black and pore phase and the subsequent reconstruction of the investigated representative volume element (RVE). Figure 6.3 presents the reconstructed microstructure of a lab-scale cathode L_1 and the commercial cathode C_1. It is obvious that the commercial electrode has a comparatively sophisticated structure comprising large agglomerates of active material particles and additional carbon fibers. This explains its high performance concerning energy and power density compared to the lab-scale electrodes. From these microstructure reconstructions it is possible to determine the volume fractions or particle diameters for all three phases and the specific interface area between them. The properties relevant to this study are listed in Table 6.2.

electrode	L_1	L_2	L_3	L_4	C_1
volume fractions / %					
LiFePO$_4$	15	22.3	36.9	38.3	56.8
CB	19.5	36.2	20.4	17.6	5.1
porosity	65.6	41.4	42.5	44.0	38.1
specific surface area / µm^2/µm^3					
a_{LFP}	4.35	7.066	9.978	8.36	7.052
a_{CB}	12.866	19.07	9.464	6.379	1.449
$a_{porosity}$	15.436	21.803	17.723	14.47	8.261
specific active surface area / µm^2/µm^3					
a_{active}	3.46	4.9	9.12	8.23	6.93

Table 6.2: Microstructure parameters of the LiFePO$_4$ cathode structures determined by SEM/FIB tomography.

The most significant trends are briefly introduced in the following. The porosity is reduced from about 65 % down to almost 40 % by the calendering step. This value is very close to the 38 % porosity of the commercial electrode. Furthermore, the volume-specific surface areas rise as the material content per volume element has been increased due to the compression. The active surface area which can be used for lithium intercalation is extended accordingly. It can be calculated from the specific surface areas by the following equation:

$$a_{active} = \left(a_{LFP} + a_{porosity} - a_{CB}\right)/2 \tag{6.1}$$

The largest active surface area among the lab-scale electrodes results for L_3 and L_4 which have been calendered and contain the largest mass fraction of LiFePO$_4$.

Two kinds of cell configuration were assembled to run the experiments in this chapter. LiFePO$_4$ half-cells similar to those in Chapter 5 were assembled in order to conduct the discharge experiments. These cells were first cycled for ten cycles between 2.7 V and 3.7 V with a 1C-rate theoretically calculated from the electrode mass. Afterwards, the cells were discharged with 1C, 2C, 3C, 4C, 5C and 10C in order to check their rate capability. Five cycles using 1C and a last cycle with C/40 have been measured subsequently. The charging before each of these steps obeys a CCCV protocol with 1C charging and C/10 cut-off current. The discharge capacities obtained for L_1 and L_4 electrodes are presented in Figure 6.4. The uncompressed cathode L_1 exhibits a strong aging due to the high discharge rates which occurs at the subsequent 1C cycles. In contrast, the calendered electrodes, for instance L_4, provide a stable capacity even after the high rate discharge cycles. A similar behavior was observed in Ref. [148]. In that study it is explained by a more elastic than plastic deformation for compressed electrodes leading to a higher stability during charge discharge experiments. The aging of uncompressed electrodes impedes in the C-rate evaluation.

Another remarkable characteristic appears for the calendered electrode at the high discharge rates in Figure 6.4b. The capacity provided decreases linearly for C-rates up to 5C because of the linear overall losses. However, it drops drastically for 10C discharge, indicating another underlying mechanism. One possible explanation would be the lithium anode which has a limited surface area. Due to high current rates it is possible that insufficient lithium can be dissolved and the overpotential increases drastically. Another reason would be a concentration gradient inside the electrolyte due to the large separator thickness compared to commercially applied separators. A similar observation was presented in Ref. [160] and explained by the same mechanisms. This limits the applicability of experimental cells for the determination of electrode performances. However, as illustrated in Figure 6.4b it is possible to investigate lab-scale electrodes with small enough capacities without having an impact of the second mechanisms. A stronger drop in discharge capacity has to be considered as an artifact of the experimental cell setup.

Energy and power density were calculated from these discharge experiments applying Equations 2.10 and 2.11 and visualized in Ragone plots.

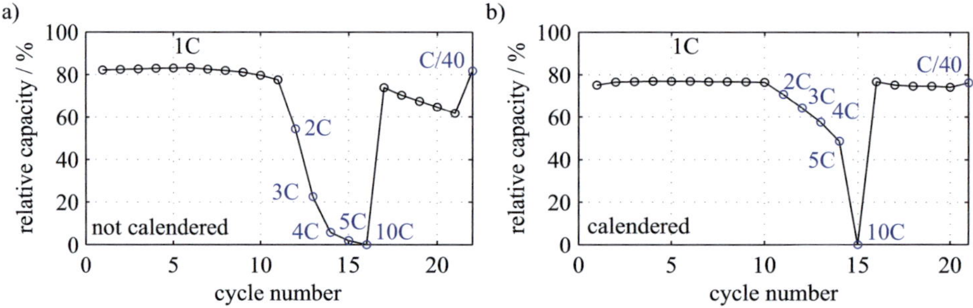

Figure 6.4: Capacities obtained for a) an uncompressed cathode L_1 and b) a compressed cathode L_4. A significant aging occurs for the not calendered electrode, whereas the calendered electrode shows stable capacities.

A second cell setup was used for the investigation of the loss mechanisms in the $LiFePO_4$-cathodes. As discussed in the previous chapter it is complicated to describe the impedance spectra of lithium anodes reliably. Therefore, a reference electrode geometry was developed in order to investigate the cathodes without effects of the counter electrode. A steel mesh was applied in the following experiments in order to examine only the cathode impedance. Such a mesh offers a reliable geometry for EIS without artifacts. However, it provides no stable potential which causes a decreasing measurement data quality for low frequencies (presented in Section 4.5) and necessitates a more intensive supervision of the quality of the data measured. Two glass fiber separators were put between the reference electrode and each main electrode in order so separate them mechanically. The electrolyte was slightly adapted in order to increase the stability of the test cells. For that, a 1M mixture of $LiClO_4$:$LiPF_6$ (9:1), in the same solvent introduced above, was applied. All impedance measurements were performed in a frequency range from 1 MHz to 10 mHz with an stimulation of 10 mV rms.

6.2 Impact of Cathode Structure on the Impedance Spectrum

Impedance spectra for all investigated $LiFePO_4$-cathodes measured via reference electrode at T=23 °C and SOC=100 % are presented in Figure 6.5. A strong variation in polarization and shape of the impedance spectra is observed among the different cathode layers. The commercial cathode C_1 reveals the smallest whereas L_1 causes the largest polarization. Furthermore, a different relation between the two characteristic semicircles in each impedance spectrum is obvious.

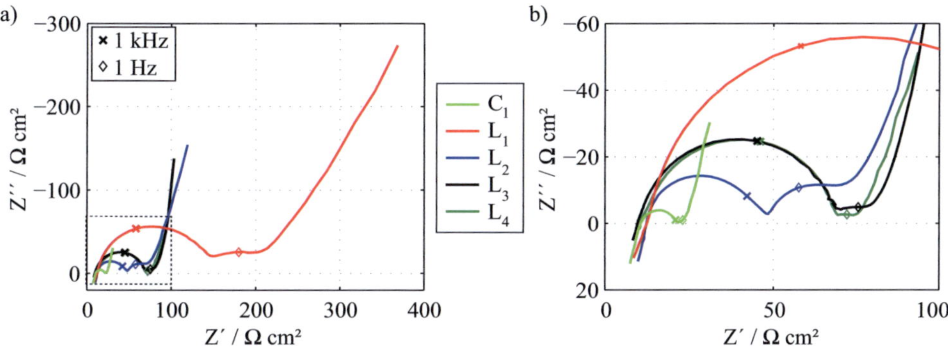

Figure 6.5: a) Nyquist plot of impedance spectra for all types of electrode measured at T=23 °C and SOC=100 %. b) Detail of the high frequency region and the commercial electrode.

The DRTs directly calculated from these spectra are presented Figure 6.6. According to Section 5.3, it is possible to evaluate the high frequency part of the DRT without a previous pre-processing of the impedance spectrum. Therefore, it is possible to analyze the identified loss process P_{2C} from Figure 6.6. It shows a strong dependence on the electrode structure, varying likewise in polarization and time constant. The previously described impedance model is adjusted and applied in the following section in order to evaluate the dependence of each loss mechanism separately and in more detail.

6.3 Adjustment of the Impedance Model

The impedance model in Chapter 5 was developed in order to describe $LiFePO_4$ half-cells. In contrast, the measurements in this chapter are conducted via reference electrode, so the impedance model has to be adjusted. It contains the following elements: One RQ-element to describe the charge transfer process P_{1C}, one RQ-element for the contact resistance P_{2C} and a 1D GFLW with serial capacity to represent the solid

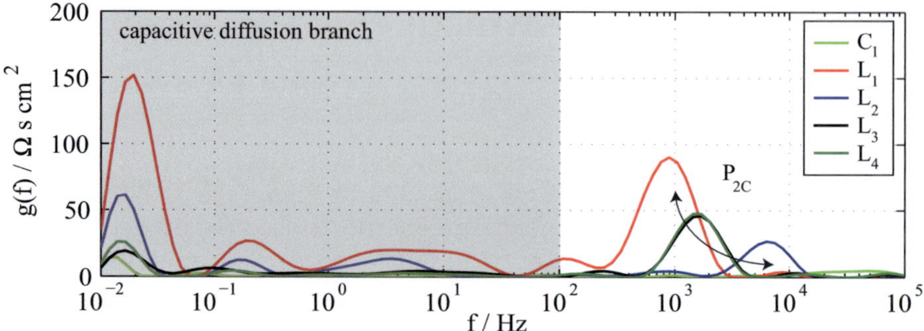

Figure 6.6: DRT of the impedance spectra for all cathode types measured at T=23 °C and SOC=100 % (P_{2C}: Contact resistance cathode/current collector).

state diffusion in the active material. Furthermore, a serial ohmic resistance and an inductance are necessary. This slightly adapted model is applied for the description of the lab-scale electrodes. Another adjustment is done for the commercial electrodes as their microstructure is fundamentally different. The commercial cathode is dominated by spherical active material agglomerates consisting of small primary particles as already shown above. Therefore, a three-dimensional diffusion into these agglomerates is assumed as solid state diffusion and accordingly, a 3D GFSW is applied as equivalent circuit element.

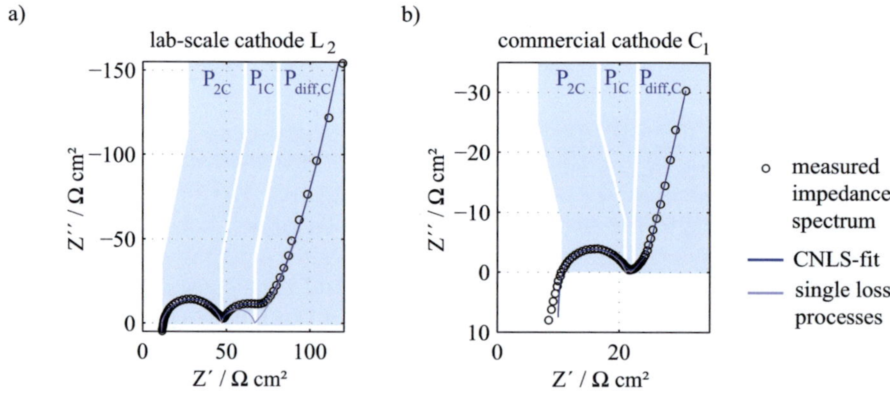

Figure 6.7: Nyquist plot of a) lab-scale cathode L_2 and b) commercial cathode C_1 with corresponding model fit and separately simulated loss processes (P_{1C}: Charge transfer cathode/electrolyte, P_{2C}: Contact resistance cathode/current collector, $P_{diff,C}$: Solid state diffusion in the active material).

Figure 6.7 compares impedance and fit result of lab-scale and commercial electrodes L_2 and C_1. The shape of the spectra is fundamentally different. Both reveal unambiguously a contact resistance P_{2C} which is represented by the high frequency semicircle. However, the contribution of charge transfer P_{1C} differs strongly. It is very dominant for L_2, whereas it almost disappears for C_1. However, the fits are very good as illustrated by the fit residuals in Figure 6.8 and they give a good estimation of polarization of P_{1C} in the commercial cathode C_1.

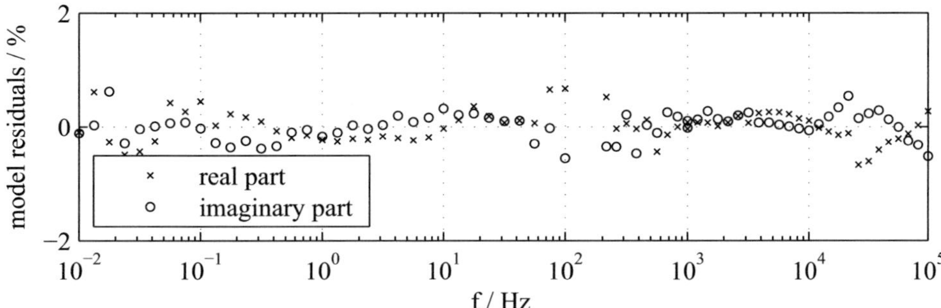

Figure 6.8: Residuals for the fit of the commercially available electrode C_1 in Figure 6.7 using the adjusted impedance model.

6.4 Impact of Cathode Structure on Loss Processes

The impedance models previously introduced were subsequently applied for the evaluation of SOC-dependent impedance spectra for each cathode layer at T=23 °C. Figure 6.9 displays the SOC-dependence of contact resistance P_{2C}. It is constant for all cathodes except Cathode L_4 which increases. This increase is caused by an aging of the contact resistance as the impedance spectra were measured from high to low SOC and the repeat measurement of SOC=100 % also showed the increase for L_4. Here the adhesion on the current collector seems to be worse compared to the other electrodes. The independence from SOC represented by all other cathodes is expected according to the previous LiFePO$_4$ evaluations in Section 5. A distinct difference in polarization caused by the fabrication can be observed among the lab-scale electrodes.

Figure 6.10 explains this schematically. Cathode L_1 is not compressed after drying, which causes a very bad adhesion to the current collector. This results in the largest contact resistance among the cathodes. Calendering (Compression) reduces the contact resistance drastically, resulting in the smallest contact resistance among the lab-scale cathodes for L_2. Next, the carbon black content is reduced for the fabrication of L_3. This leads to an increase of P_{2C} as the carbon black improves the electrical contact to the current collector. The last step of doubling the layer thickness to obtain L_4 does not influence P_{2C}.
These dependencies are very clear and easy to explain. They prove the physical interpretation and show that the tracking of P_{2C} for electrode variation is enabled by

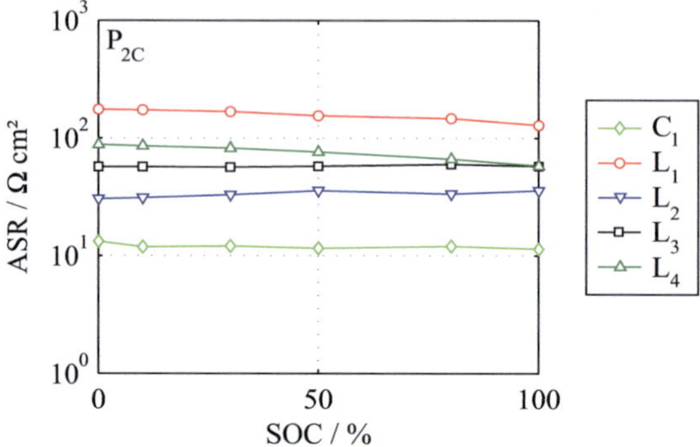

Figure 6.9: Illustration of contact resistance P_{2C} for all investigated electrodes and SOC at T=23 °C.

the impedance model. Finally, the commercial cathode reveals the smallest contact resistance among all electrodes, indicating an improved fabrication. Possible advancements can be achieved by stronger compression or a pre-treatment of the current collector. Furthermore, the coating process itself may influence the contact due to an optimized carbon black or binder distribution.

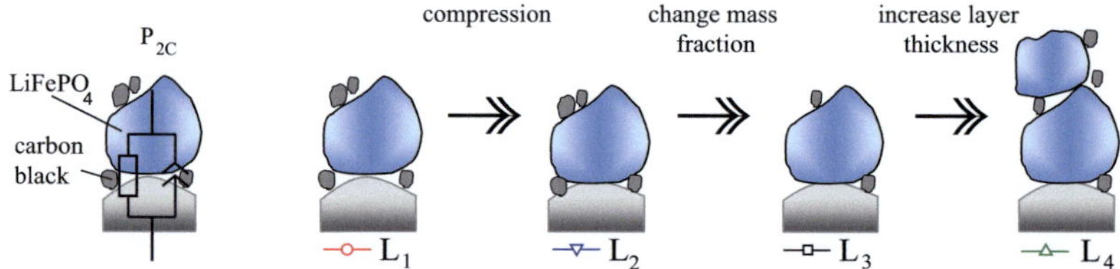

Figure 6.10: Schematic of the contact resistance (P_{2C}) dependence on electrode fabrication and microstructure changed from L_1 to L_4 by 1. compression, 2. increase of active material fraction and 3. increase of layer thickness.

The charge transfer P_{1C} presented in Figure 6.11 exhibits a pronounced SOC dependence, revealing an increase of polarization with decreasing SOC. It increases strongly in the upper SOC region of L_2, L_3 and L_4, whereas the increase reduces for low SOC. L_1 is slightly different concerning its charge transfer dependency, also having the highest polarization. The polarization of P_{1C} is steadily reduced from L_1 to L_4. This reduction among the lab-scale electrodes will be discussed in the following section with the help of microstructure parameters. P_{1C} is smaller for the commercially used compared to the lab-scale cathodes and increases constantly with reduced SOC. The solid state diffusion shows the same behavior for all cathodes. It increases steadily with decreasing SOC, except for SOC=100 %. Its polarization is the lowest for the commercial, whereas it varies among the lab-scale cathodes. These variations must be explained by the microstructure, as the intrinsic material properties are similar for all cathodes. The charge transfer and the solid state diffusion will be discussed in the following section with respect to the microstructure parameters.

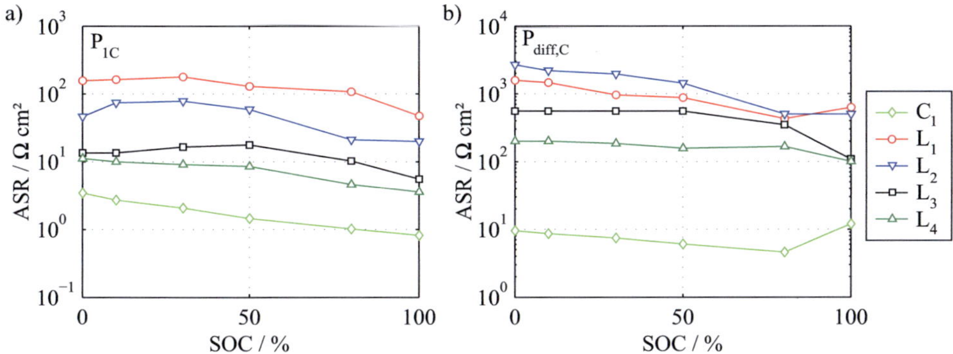

Figure 6.11: Polarization of a) charge transfer P_{1C} and b) solid state diffusion $P_{diff,C}$ for all investigated electrodes and SOC at T=23 °C. The ASR is normalized by the electrode area.

6.5 Correlation with Microstructure

A calculation of the specific surface area between active material and electrolyte is enabled by the microstructure reconstructions introduced above. With this and the cathode layer thickness it is possible to determine the active surface area for each cathode investigated. In this section, the ASR of charge transfer and solid state diffusion is not normalized by the electrode area as is usually done, but rather it is normalized by the active surface area determined by reconstruction. Therefore, it is denominated as active area specific resistance (aASR) here and in the following chapters. The resulting values for P_{1C} and $P_{diff,C}$ are displayed in Figure 6.12. It is obvious that the polarization of different cathode structures converges when applying the active surface area. Especially for L_2 to L_4, the polarizations of P_{1C} and $P_{diff,C}$ are very similar, indicating that only the microstructure determines their polarization. The polarization of charge transfer is still larger for L_1, indicating another impact. As L_1 was fabricated in a second batch compared to all other lab-scale cathodes, it is possible that a different microstructure emerged. For instance, a deviant binder distribution or a less homogeneous microstructure might have resulted, leading to an impeded charge transfer. The slight variation among the other lab-scale cathodes might also be caused by an inhomogeneous microstructure, an influence of the model fit or binder distributions which cannot be resolved by the reconstruction.
A drastically smaller value for both polarization processes is obtained for the commercial electrode. A part of this difference is explained by the microstructure. The larger

active surface area due to the high specific surface area and the thick layer allows for a superior charge transfer. Therefore, the correct normalization leading to the results in Figure 6.12 approaches the commercial and the lab-scale electrodes. The remaining difference may depend on the active material itself. However, nothing definite can be said about this as the exact active material is not known. It could be different because of intrinsic material properties, the outer shell treated by the carbon coating or the influence of binder on the surface. Furthermore, the microstructure reconstruction of the commercial cathode contains several uncertainties as the micro porosity inside the agglomerates cannot be determined unambiguously.

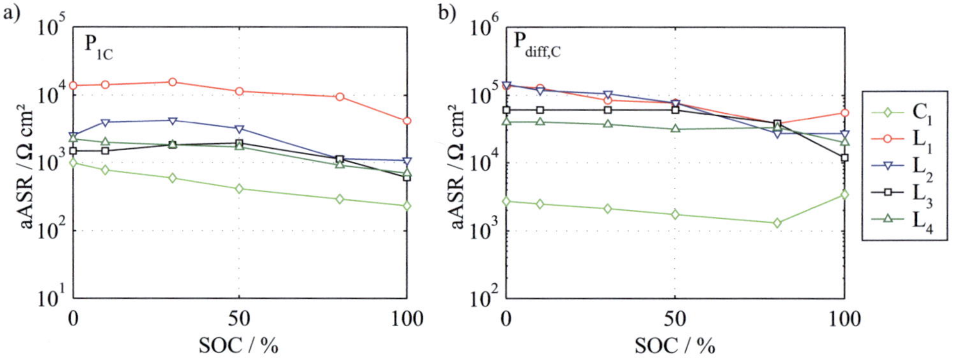

Figure 6.12: Polarization of a) charge transfer P_{1C} and b) solid state diffusion $P_{diff,C}$ for all investigated electrodes and SOC at T=23 °C. The aASR is normalized by the active surface area determined by microstructure reconstruction.

To sum up, some basic statements on the influence of electrode fabrication on the electrochemical loss mechanisms can be given:

1. The contact resistance P_{2C} can be minimized by calendering the electrodes and by using a high carbon black content.
2. Charge transfer P_{1C} and solid state diffusion $P_{diff,C}$ are directly proportional to the active surface area between active material and electrolyte.
3. The difference among commercial and lab-scale cathodes' impedance can partly be explained by the microstructure. A remaining difference may result from different active materials or material distribution.

A basic dependence of loss mechanisms on microstructure was given in this section. Next, the influence of these loss mechanisms on the electrode performance under load is analyzed.

6.6 Impact of Cathode Structure on Performance

In this section, the electrode performance under load is compared by running discharge experiments with different C-rates and plotting the resulting Ragone diagrams. Applying this, the performance improvement from L_1 to L_4 is presented and discussed from the aspect of the loss processes previously identified. The first step in order to improve the cathode performance was a calendering step and therefore a reduction of the contact resistance P_{2C} and the charge transfer P_{1C}. Figure 6.13a illustrates that this step allows for a much better performance for high discharge rates. This improvement can be caused by the two reduced loss mechanisms P_{1C} and P_{2C} but also by an enhanced electronic layer conductivity achieved by a better percolation of the electronic paths. The impact of a reduced carbon black content is presented in Figure 6.13b. A higher energy density can be achieved for low to medium discharge rates as a diminished mass fraction of inactive material is used. The influence of higher contact resistance and decreased electronic layer conductivity is not significant. However, for higher rates these losses become more significant and the energy density reductions become more pronounced compared to the cathodes with high carbon black content. The assignment of this stronger reduction to the electronic path and the contact resistance is unambiguous as the porosity remains similar and therefore prevents a change of the losses inside the electrolyte. Furthermore, the charge transfer resistance is reduced from L_2 to L_3, proving that the charge transfer is not responsible for the high rate performance limit.

The reduced electronic layer conductivity was not identified by the impedance analysis which indicates that the electronic conductivity itself does not change drastically. However, another effect related with the electron transport can be the cause for a performance drop at high discharge rates. A bad percolation of the electronic paths may lead to electron depletion causing large overpotentials for high discharge rates. This effect cannot be identified by impedance spectroscopy because it only occurs at large currents.

In a last step, the layer thickness was increased by a factor of two. Naturally this does not, as presented above, have an impact on the contact resistance, but it does reduce the absolute charge transfer polarization. The latter remains constant if it is normalized by the active surface area. Nevertheless, the increasing thickness causes a reduction of energy and power density. This correlation indicates that the limited electronic layer conductivity causes the reduction.

Summing up, the influence of material fractions and calendering on the electrode performance leads to the following conclusions. Both parameters affect the performance at low and high discharge rates differently. An increase of energy density at low discharge rates is achieved by reducing the carbon black content. This is an obvious correlation as, in this case, less inactive components are part of the composite layer. For high discharge rates this has a counterproductive effect.

The electronic layer conductivity and the contact resistance between cathode and current collector are identified as dominating loss mechanisms, influencing the high rate performance. The charge transfer does not have an influence, although it reveals comparable values for polarization. Similarly, the solid state diffusion and the lithium-ion transport in the porosity do not have an effect for the electrodes investigated.

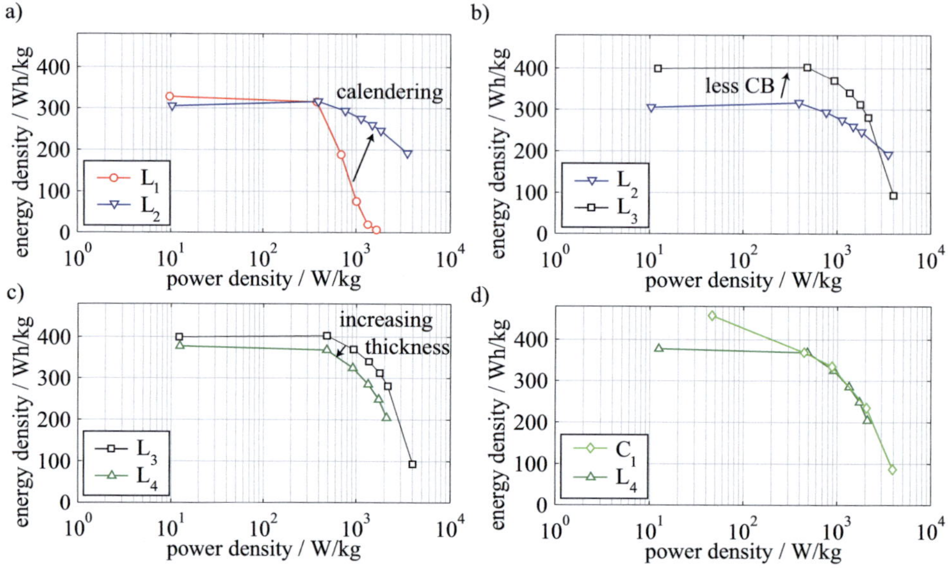

Figure 6.13: Ragone diagrams for lab-scale and commercial cathodes, referred to the pure cathode layer mass without the current collector. Subfigures a) to d) show a comparison of the electrodes.

Figure 6.13d presents the comparison of lab-scale cathode L_4 and commercial cathode C_1. The performance at high discharge rates is similar, providing the same energy and power densities referred to the pure cathode mass. For low C-rates, the high carbon black content and the low exploitation of active material for small currents cause an inferior performance of the lab-scale cathode. Yet, it is necessary to compare the energy densities referred to the electrode mass together with the current collector in order to obtain a fair comparison of the electrode performance in a real cell. The mass of the current collector was taken to be half of the commercially used as both sides are usually coated in commercial cells.

The obtained Ragone diagram is presented in Figure 6.14. A clear advancement is achieved with every step of cathode adaption. However, even the best lab-scale cathode structure L_4 reveals a lower performance compared to the commercial cathode. This difference would be even larger if all other inactive components, for instance electrolyte, separator or cell case would be considered as additional inactive mass. According to the previous analysis this superior performance of the commercial cathode indicates that its electronic layer conductivity is much larger, although the latter has a significantly lower carbon black content. Several factors can lead to this superior performance. First, a better mixing of the wet cathode slurry can produce an improved percolation of electronic paths. Second, additional carbon fibers, as found in the commercial cathode layer, may provide a faster electron transport. Then, the formation of agglomerates or a higher compression may also allow for a superior connectivity of electronic paths, leading to a better cathode performance.

6.7 Discussion and Conclusions

Various lab-scale cathodes with different material fraction, layer thickness and compression have been analyzed in this chapter. First, electrochemical impedance spectroscopy and impedance models were applied in order to quantify the contribution of each loss processes and to correlate their polarization with microstructure parameters. The contact resistance P_{2C} could be optimized by compressing the electrodes and by increasing the carbon black content. The charge transfer process P_{1C} mainly depends on the active surface area between cathode and electrolyte. It is impeded by a high carbon black content as then, less active surface area is provided. Furthermore, it naturally decreases with increasing layer thickness. The solid state diffusion was not investigated systematically as the particle size could not be varied. The dependency of loss mechanisms on electrode structure proved the physical interpretation introduced in Chapter 5. All occurring loss mechanisms are larger for the lab-scale cathodes compared to the commercial one.

Second, the performance of these cathodes was analyzed by discharge experiments. By comparing these results and the impedance analysis it was possible to identify contact resistance P_{2C} and electron transport through the cathode layer as main contributors to the limited performance at high rates. The latter cannot be identified by the impedance analysis as the electronic conductivity itself is not crucial. However, electron depletion may occur at current collector contact or more probable at inter-particle contact only for higher discharge rates. Consequently, these constrictions lead to large overpotentials. Neither charge transfer or solid state diffusion, nor lithium diffusion in the electrolyte are a deciding factor for the investigated electrodes.

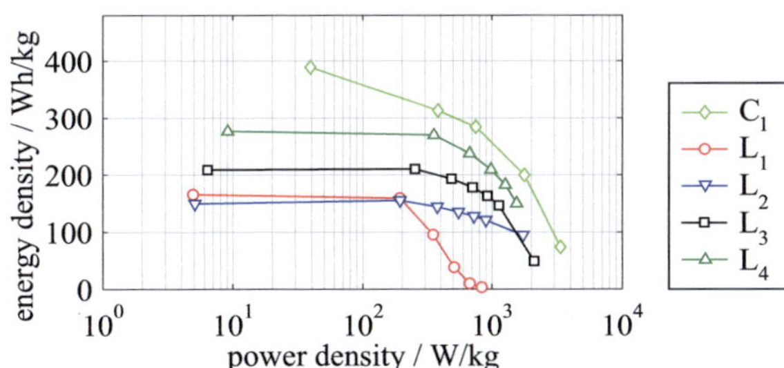

Figure 6.14: Ragone diagram for lab-scale and commercial cathodes, referred to the sum of cathode layer and current collector mass.

These conclusions hold true for the specific cathode structure; however, they cannot be generalized to other electrodes. For instance the lithium transport in the pores would play a major role for a lower porosity. Therefore, an optimum of high porosity and good percolation of the electronic path must be found for every specific electrode composition. The commercial cathode, compared to the lab-scale cathodes in this study, demonstrated a better percolation by an additional use of carbon fibers. This results in an improved layer conductivity and allows for a much higher layer thickness without a limitation of the high rate performance.

Several studies about the optimization of microstructures concerning these parameters have been published. In Ref. [161], a porosity of slightly below 40 % is determined by conductivity measurements and simulations as optimal trade-off between good electronic and ionic conductivity. However, no discharge experiments with the introduced electrodes in order to prove the theoretical findings are presented. Tran et al. [148] identify the electronic path including the contact resistance as major contribution to the performance limit of their electrodes by discharge experiments, which is in accordance to this study. Other studies such as Refs. [162, 163] found the ionic conductivity in the electrolyte as the main factor. A correlation of electrochemical loss processes identified by impedance spectroscopy and discharge experiments has not been published among these studies. In contrast

This combination is a profitable approach and it reveals that not all significant contributions can be identified via impedance spectroscopy. A combination of impedance spectroscopy, discharge experiments and optionally supplemental measurements of the layer conductivity yields a powerful method to optimize electrode structures. Comparing the published studies and the results obtained here leads to the conclusion that the limitations of the electrode performance depend on the applied active material, the material fractions, the layer thickness and additional factors. A systematic optimization can be conducted by the approach previously introduced combining discharge experiments and impedance analysis. The results obtained for the cathodes analyzed in this study cannot be generalized but the approach applied can be transferred to any electrode structure.

7 Modelling of Complex Anode Structures

In this chapter, the characterization of graphite-anodes extracted from commercially available lithium-ion cells is presented. The obtained results will be transferred to the commercial cell in the last part of this thesis. The anodes are analyzed by electrochemical impedance spectroscopy and afterwards evaluated using the DRT. The resulting DRTs are correlated with electrochemical loss processes and their interaction with the underlying microstructure. The integration of this interaction into the impedance model is introduced in order to obtain a physically meaningful evaluation of graphite impedance spectra. Thereby, not only the active surface area as in the previous chapter but also the impact of ionic and electronic conductivities on the impedance spectrum are analyzed.

This chapter is intended to clarify the following points:

1. Are the previously introduced methods suitable to analyze commercially used graphite-anodes?
2. Is the use of simplified serial impedance models sufficient for the analysis of more complex electrodes?
3. Which are the loss mechanisms dominating a graphite-anode?

Parts of the measurements presented in this chapter have been published in Ref. [60].

7.1 Electrodes and Measurement Setup

Graphite-anodes were extracted from the commercial high power 18650 cell which will be introduced in Chapter 8. The extraction and preparation procedure needed to obtain suitable graphite-anodes will also be presented there. One side coated graphite-anodes with a layer thickness of 48 µm on a copper current collector turn out. These anodes were reconstructed by Moses Ender at IWE applying micro X-ray tomography [164] and a subsequent evaluation as introduced in Refs. [147, 159] in order to determine the microstructure parameters. A porosity of 35.2 % and a tortuosity of 2.72 were calculated. Furthermore, an average particle size of 4.236 µm was determined which is much larger compared to the $LiFePO_4$-cathodes previously introduced.

The graphite-anodes were measured in full cell configuration with the corresponding $LiFePO_4$-cathodes also extracted from the commercial cell. This allows for a direct comparison of experimental and commercial cell results in the next chapter. The electrodes

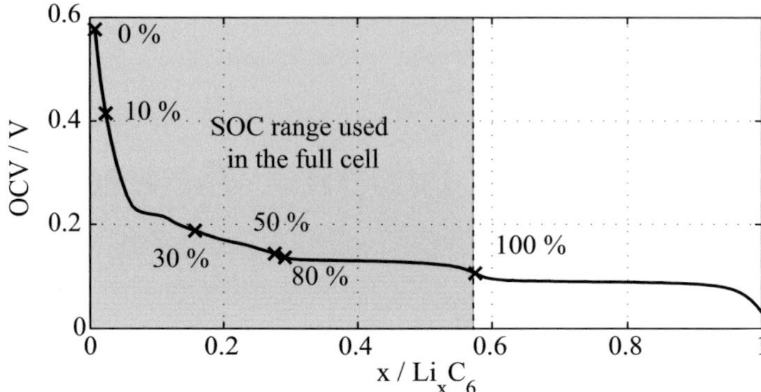

Figure 7.1: OCV curve of a graphite-anode, measured in half-cell configuration. The SOC which correspond to the full cell are marked.

were assembled in EL-Cell housings [135] with an electrode area of $2.54\,\mathrm{cm}^2$ using a steel mesh as reference electrode. Two glass fiber separators of $220\,\mu\mathrm{m}$ thickness were put between reference and each main electrode in order to separate them mechanically. A $1\,\mathrm{M}$ mixture of $LiClO_4{:}LiPF_6$ (9:1) in EC:EMC (1:1) solvent was used as electrolyte. After assembly, the full cells were cycled between 2.0 and 3.6 V for twenty times with a current of $3.5\,\mathrm{mA}$ obeying a CCCV protocol. Then, impedance spectra were measured under SOC variation (100 %, 80 %, 50 %, 30 %, 10 %, 0 %). The SOC was stepwise decreased from 100 % to 0 %, whereas it was defined by the integration of discharge current and the previously measured discharge capacity. The open circuit potentials of the graphite-anode which correspond to these SOCs could be determined by additional OCV measurements of a full cell using a $LiTi_5O_{12}$ reference electrode. These SOC points have been marked in a graphite OCV curve presented in Figure 7.1 in order to demonstrate the SOC range investigated. It is obvious that not the entire graphite-anode is used in the full cell. This phenomenon is known as electrode matching and it was explained in Section 2.1.5.5. Furthermore, the close position of SOC=80 % and 50 % is remarkable. This can be caused by slightly different OCV of the full and half-cell setups compared here, leading to an apparent shift of SOC=80 % to 50 %. In this chapter, the SOCs are given by the discharge capacity of the full cell. Afterwards, impedance spectra were measured for varying temperatures (30 °C, 23 °C, 20 °C, 10 °C, 0 °C) at three different SOCs (100 %, 50 %, 0 %). All impedance measurements were performed in a frequency range from $1\,\mathrm{MHz}$ to $10\,\mathrm{mHz}$ with an stimulation of $10\,\mathrm{mV}$ rms for high and $5\,\mathrm{mV}$ rms for low frequencies (f<100 Hz).

7.2 Impedance Spectrum of Graphite-Anodes

A Nyquist plot of the graphite-anode impedance at SOC=100 % and T=23 °C and the corresponding DRT are presented in Figure 7.2. The impedance spectrum reveals two semicircles, one small semicircle at high and a large and very flat one at medium frequencies. No additional loss process is indicated behind the characteristic bend, assuming merely the solid state diffusion to occur at these frequencies. As a consequence, the DRT is only evaluated for higher frequencies which corresponds to the first procedure introduced in Section 5.3 for applying the DRT on lithium-ion batteries.

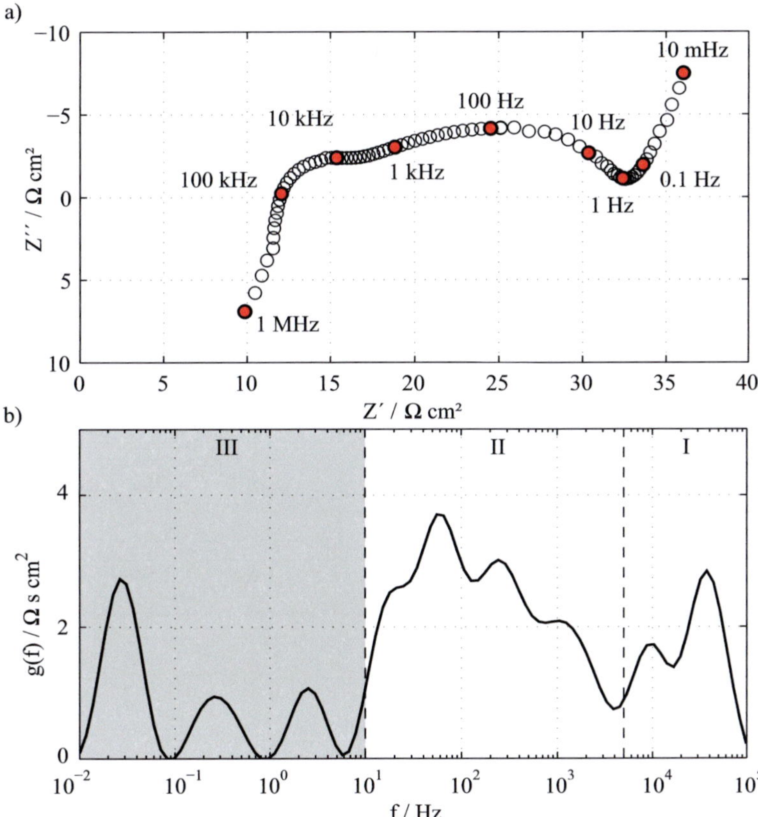

Figure 7.2: a) Nyquist plot of impedance spectrum and b) corresponding DRT of a graphite-anode at 100 % SOC and T=23 °C.

The measured frequencies are divided into three characteristic frequency ranges, denominated as I, II, and III. Frequency range I includes the high frequencies between

5 kHz and 100 kHz. Here, two characteristic peaks are distinguishable in the DRT of
the graphite-anode, whereof the small peak can be a side effect of the large semicircle in
range II or the change of the measurement range occurring at 20 kHz. An unambiguous
identification of two physical loss processes is therefore not possible. Frequency range
II covers frequencies between 10 Hz and 5 kHz. The DRT indicates the existence of up
to four characteristic time constants, which is a first hint of either four individual loss
processes or less, if one loss process with several coupled time constants takes place
here (see Section 2.4). Frequency range III, assigned to frequencies below 10 Hz is not
analyzed by the DRT in this section. It is dominated by the capacitive behavior and
the solid state diffusion.

7.3 Impact of SOC and Temperature

Next, the SOC variation between 100 % and 0 % SOC at T=23 °C is considered, displayed
in Figure 7.3a. The anode impedance does not change strongly for SOCs above 10 %.

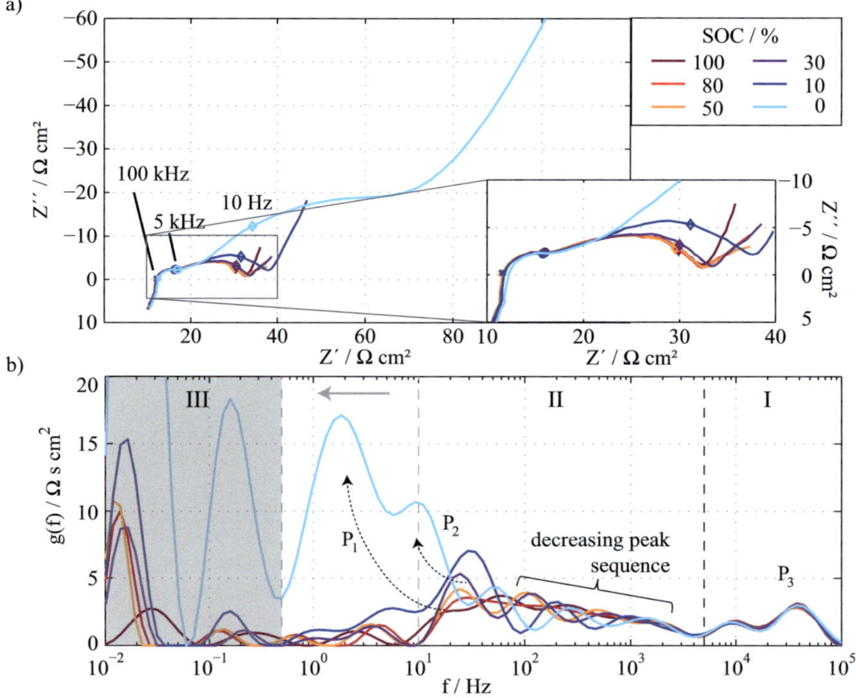

Figure 7.3: a) Nyquist plot and b) DRT of graphite-anode impedance spectra for
varying SOC at T=23 °C.

The ohmic resistance and the first semicircle at higher frequencies do not show any SOC-dependency at all. The impedance only shows a pronounced SOC-dependency for low frequencies and for very low SOC (10 % and below). Figure 7.3b presents the corresponding DRTs of the anode impedance for varying SOC. There are small variations in the shape of the calculated DRTs for SOC above 10 % within range II, whereas range I remains totally constant. The DRT changes drastically in range II at SOC=0 %. Here, the predominant peak P_1 increases significantly in height and its maximum shifts to a frequency more than one order of magnitude lower. The second predominant peak P_2 and the decreasing peak sequence for higher frequencies in frequency range II possesses a weak SOC-dependency. Note that the lower border of frequency range II was shifted from f=10 Hz to f=0.5 Hz in order to include the SOC dependent peak.

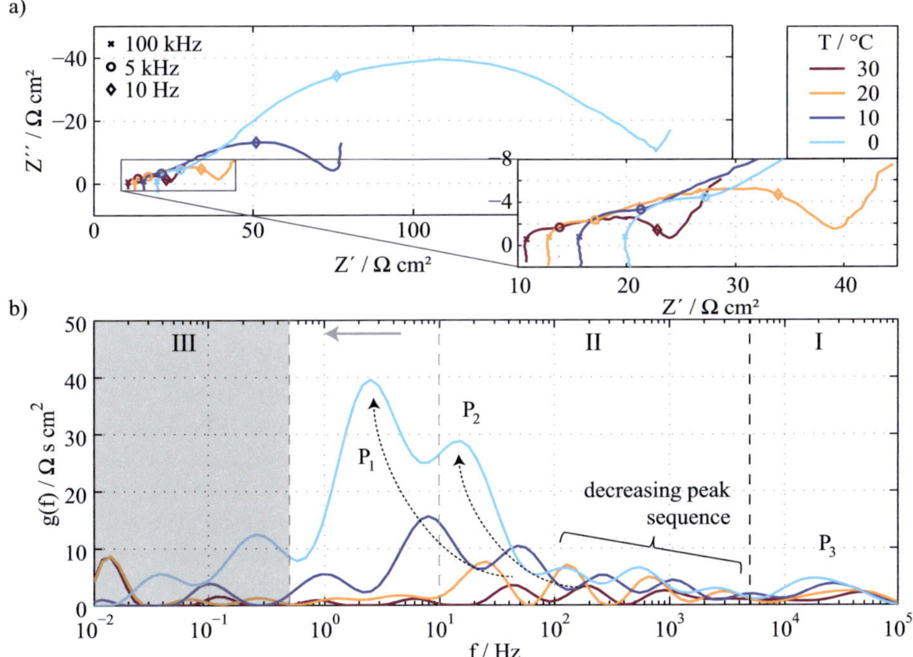

Figure 7.4: a) Nyquist plot and b) DRT of graphite-anode impedance spectra under temperature variation at SOC=100 %.

Figure 7.4 displays the graphite impedance spectra and the corresponding DRTs at SOC=100 % for varying temperatures. The impedance spectra show a strong raise of polarization and a slight increase of the ohmic resistance. The DRT allows for a more detailed analysis. A weak temperature dependency can be observed in frequency range I. The DRT increases slightly for decreasing temperatures, then revealing only one characteristic peak, denominated as P_3 in the following analysis. Once more, frequency range II shows the most distinct parameter-dependency. Both main peaks and the peak sequence at higher frequencies increase steadily with decreasing temperature. A clear identification of two main peaks P_1 and P_2 is possible.

7.4 Equivalent Circuit Elements and Model Structures

The DRT calculated from the anode impedance spectrum (see Figure 7.2) shows a very complex structure represented by a broad peak distribution. Frequency range I and II already contain six peaks, which could be assigned to six electrochemical loss processes. However, the use of six independent impedance elements would not only cause a very unstable fit but would also neglect that a higher number of peaks can be caused by more complex loss processes, by showing for instance Warburg behavior (see Section 2.4.1.4) or more complex model structures as TLM (see Section 2.4.3). As described in Chapter 5 using a minimal amount of impedance elements is recommended because they allow for a satisfactory description of the impedance spectrum and provide a good physical interpretation with the most meaningful fit results. The information from DRT evaluation can be used to receive information about reasonable model structures and impedance elements.

In this section, different impedance models and model structures are introduced and compared with regard to their physical basis as well as their fit quality and difference of fitting results. First of all, possible physical interpretations and modelling approaches are discussed for each frequency range.

Frequency range I shows one large peak P_3 at 40 kHz and a small peak for lower frequencies. These peaks do not change for varying SOC. For decreasing temperature, the polarization increases slightly and only one peak dominates. The high characteristic frequency and the SOC-independence indicate that this loss process could be caused by contact resistances between graphite and current collector or inter-particles contact resistances. Therefore, one RQ-element is used in order to represent the underlying loss process. The small peak is not taken into account as it could be an artifact of switching the measurement range and it does not contribute significantly.

The DRT reveals four characteristic peaks in frequency range II. A better understanding of their interrelation can be received from parameter variations. For decreasing SOC, P_1 shows a strong increase of polarization and characteristic time constant. A second peak, P_2, shows a weak SOC-dependency and the following peak sequence does not increase significantly. Both main peaks show a strong temperature dependency and also the following peak sequence increases steadily for decreasing temperatures. This leads to several alternatives for the impedance model design.

First of all, it is assumed that two main loss mechanisms cause the broad peak distribution in frequency range II. These loss processes are assumed to be charge transfer at graphite surface and transport through the SEI. The peak sequence for higher frequencies is expected to be a result of side peaks of these main loss processes or the electrode structure. Two impedance elements are used to model charge transfer and SEI in frequency range II. An RQ-element seems to be most suitable for charge transfer as this is a typical application of RQ-elements. An RQ-element can be used for transport through the SEI as well. However, there are more complex impedance elements which are able to describe the SEI appropriately. One candidate for SEI-modelling is the so called HN-element (introduced in Section 2.4.1.3). The HN-element was introduced to describe relaxation processes in thin polymer layers [91], which is very similar to the lithium-transport through the SEI. Furthermore, the HN-element can contribute to a decreasing peak sequence as observed at the graphite-anode for SOC- and temperature variation and is therefore also motivated by the DRTs measured.

Furthermore, solid state diffusion of lithium inside the graphite particles has to be taken into account in order to model frequency range III. Different Warburg elements can be used to describe solid state diffusion for different material structures (as explained in Section 2.4.1.4). In this study, three kinds of solid state diffusion models are compared. All three models are Finite Space Warburg-elements, which means that the capacitive behavior is already included into the solid state diffusion model. 1D-, 2D- and 3D-Warburg elements are applied in order to fit the measured impedance spectra.

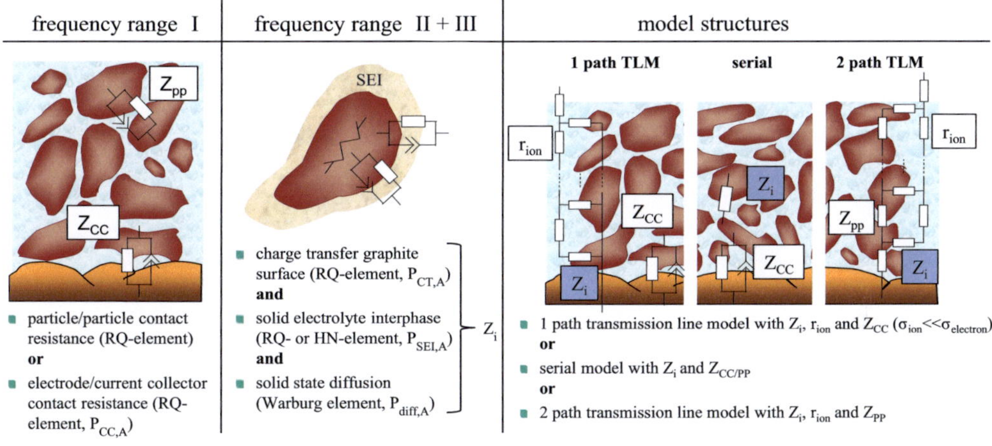

Figure 7.5: Overview of possible graphite-anode equivalent circuit model elements and different equivalent circuit model structures.

All these loss processes and their specific impedance elements can be combined into different model structures. The basic model structure is a serial connection of all impedance elements. This is very elementary and is often satisfactory in terms of accuracy and physical interpretation. However, because of the specific DRT and the broad peak distribution, in this case a combination of these impedance elements into a transmission line model could be reasonable. As introduced in Section 2.4.3, a TLM can lead to a decreasing peak sequence with small side peaks. The structure of transmission line models contains an electronic path, an ionic path and an interface resistance between these paths. In the case of graphite-anodes, the interface resistance is defined

model ID	model structure	frequency range I	frequency range II	frequency range III
1	serial model	1 RQ-element	2RQ-elements	2D GFSW
2	serial model	1 RQ-element	1RQ-/1HN- element	2D GFSW
3	1 path TLM (R_{ion} fix)	1 RQ-element	2RQ-elements	2D GFSW
4	1 path TLM (R_{ion} free)	1 RQ-element	2RQ-elements	2D GFSW
5	1 path TLM	1 RQ-element	2RQ-elements	1D GFSW
6	1 path TLM	1 RQ-element	2RQ-elements	3D GFSW
7	2 path TLM	1 RQ-element	2RQ-elements	2D GFSW

Table 7.1: List of possible graphite-anode equivalent circuit models.

by charge transfer, SEI and solid state diffusion as all these loss processes occur at the interface between electronic and ionic conduction, illustrated in Figure 7.5. Here the ionic path is given by the ionic conductivity of lithium-ions through the pores. The electron transport through the graphite-anode dominates the electronic path of the transmission line model. However, in this case there are two possible ways to describe the electronic path.

The first way is to assume that the effective electronic layer conductivity is much higher than the ionic conductivity in the electrolyte (bulk conductivity σ_{ion} =0.75 S/m), which is justified for graphite-anodes ($\sigma_{electron}$ ≥1000 S/m [103]). In this case, the electronic path can be short-circuited and the model will be denominated as one path transmission line model (1 path TLM) in the following sections. The second approach is that the high frequency loss process is caused by the inter-particle resistance. In this case, the RQ-element, which describes the high frequency loss process must be integrated into the electronic path of the transmission line model. The latter will be denominated as two path transmission line model (2 path TLM) in the following sections.

All of these possible impedance models are illustrated in Figure 7.5 in order to give a short overview. Furthermore, Table 7.1 lists all possible impedance models and assigns individual IDs to them in order to distinguish them during the following analysis.

7.5 Fitting Complex Impedance Models with Help of DRT and Microstructure Parameters.

The DRT is not only a beneficial tool for the development of physically motivated impedance models but it can also be helpful for the parametrization of impedance models and for the choice of starting parameters. As already discussed in Section 2.2.4, it can be used to choose starting parameter as polarization (area under the peak) and characteristic time constant (position of peak maximum) for simple loss processes described by RQ-elements. For more complex impedance models such as TLM, additional parameters are needed in order to achieve physically meaningful starting parameters. The equation of TLM is given in Section 2.4.3. Two parameters are needed in order to parametrize the TLM physically meaningful:

1. The electrode thickness L: It can be determined by SEM, laser microscope or micrometer caliper. The thickness of the graphite-anode investigated in this thesis was determined by SEM to be L =48 µm.

2. The differential ionic resistance r_{ion}: The resistance caused by the limited ionic conductivity inside of the porosity is defined by [52]

$$R_{ion} = \frac{1}{\sigma_{ion}} \cdot \frac{L}{A} \cdot \frac{\tau_p}{\epsilon} = r_{ion} \cdot L. \tag{7.1}$$

Using the ionic conductivity σ_{ion} =0.75 S/m, the electrode area A=2.54 cm^2, the porosity ϵ =0.35 and the tortuosity of the pores τ_p =2.72, Equation 7.1 leads to the differential ionic resistance r_{ion} =0.040 Ωµm^{-1} for the anode investigated in this chapter.

These two values, layer thickness and differential ionic resistance, define a physically meaningful parameter choice for the TLM. They can be fixed for the CNLS-fit in order to get physically meaningful results and to check whether the transmission line behavior may have an effect on the impedance spectrum of the investigated anode. Furthermore, a fit with free ionic resistance is tested in order to allow deviations from the expected values. Such deviations could be explained by additional effects inside the open porosity which cannot be determined by microstructure analysis. Those effects will be discussed in a later section. The other loss process parameters are chosen according to the DRT and normalized on electrode thickness as already explained for the transmission line model structure (see Section 2.4.3).
A meaningful fit of transmission line models is more complex for SOC- and temperature-variation. The TLM-parameters are fixed during the fit for SOC-variation as they are not expected to change significantly. The ionic resistance is either fixed at the calculated value (model 3), or it is free for SOC=100 % and remains constant for the other SOC (model 4). For temperature-variation, the ionic conductivity inside the pores is expected to change. A two step fitting procedure was therefore applied in this study:

- First fit: During the first fit, the ionic resistance inside the pores is set as a free fitting parameter beside all other model parameters. The resulting fit delivers a temperature dependency for the ohmic resistance, which represents the temperature dependency of the electrolyte inside the separator.

- Assumption for the second fit: The temperature dependency is expected to be the same for ohmic resistance and ionic conductivity inside the pores. As the fit of the ohmic resistance is very stable, its temperature dependency from the first fit is transferred to the temperature dependency of the ionic conductivity inside the pores at the second fit.

- Second fit: The ionic resistance r_{ion} is not a free fit parameter, but it follows the temperature dependency of the ohmic resistance previously determined. Using this fixed temperature dependency of ionic conductivity inside the pores, a second fit is conducted to determine the remaining model parameters.

7.6 Fit Quality and Model Evaluation

In this section, a comparison of these various impedance models concerning fit quality and fit results is conducted. A first example for a fit result is shown for model 1 (see Table 7.1) consisting of a serial connection of 1 RQ-element for frequency range I, 2 RQ-elements for frequency range II and a 2D GFLW for frequency range III. It is compared to model 3, where the same impedance elements are combined in a TLM structure. For model 3, the TLM parameters are fixed at the values calculated from microstructure parameters. Figure 7.6a displays a measured impedance spectrum and fit results for the two models. Both fits look very good and do not have a difference in their impedance curve. Figure 7.6b shows the corresponding DRTs which are also equal. In frequency range I, both models cannot represent the occurring peaks exactly. However, this is partially caused by the switch of the measurement range at 20 kHz which influences the peaks in the DRT. Furthermore, only one RQ-element is used in order not to over-interpret the occurring peak number.

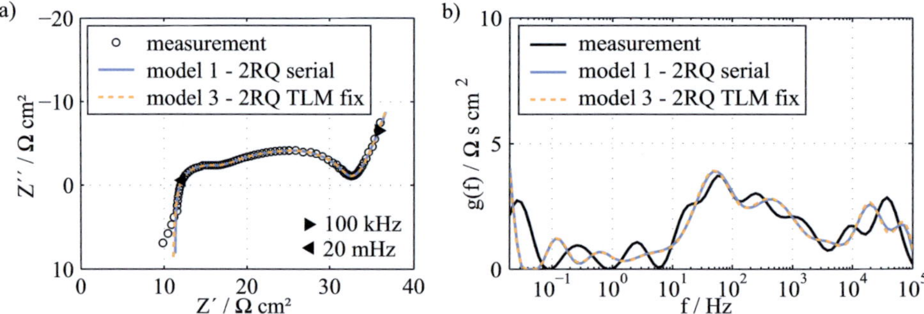

Figure 7.6: a) Impedance spectrum of graphite-anode for T=23 °C and SOC=100 % and fit of model 1 (serial) and model 3 (TLM) for the marked frequency range. b) Corresponding DRT of measured impedance spectrum and fit result.

Both models can approximate a broad peak distribution in frequency range II, however an exact representation of single peaks is not possible. It is obvious that only two main peaks dominate the model DRT whereas the impedance curve provides four peaks in this frequency range. The corresponding fit residuals in Figure 7.7 show very small deviation and remain right below 0.5 % for all frequencies. Slightly increasing residuals can be observed for the lowest frequencies and for 20 kHz where the measurement range is switched. These two fit results demonstrate the ambiguity of the proposed impedance models.

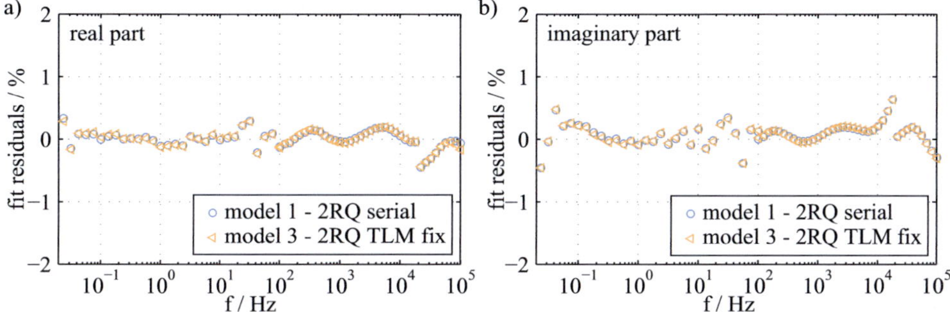

Figure 7.7: a) Real part and b) Imaginary part of the residuals approximating model 1 (serial) and 3 (TLM) to the graphite-anode impedance spectrum for T=23 °C and SOC=100 %.

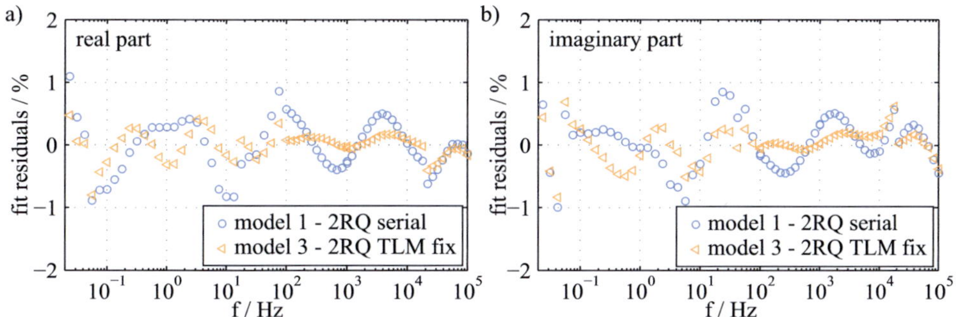

Figure 7.8: a) Real part and b) Imaginary part of the residuals approximating model 1 (serial) and 3 (TLM) to the graphite-anode impedance spectrum for T=23 °C and SOC=0 %.

Figure 7.8 shows the residuals for both models for SOC=0 % at 23 °C. It is obvious that model 3 with the TLM structure provides a better fit concerning real part and imaginary part of the residuals. Especially frequency range II shows much lower residuals for the TLM as therein, the TLM behavior is dominant in the impedance spectrum and hence in the DRT. This difference demonstrates the importance of a comparison for all operating conditions measured.

Because comparing the relative residuals of all the models and fits is confusing, another measure must be used in order to compare the fit quality of this high number of fits. The RMSE (as introduced in Section 2.2.4.2) is calculated for each fit in the specified frequency range. In this section, frequencies between 20 mHz and 100 kHz were selected in order to fit the impedance models and to calculate their fit quality. For lower frequencies, the Kramers Kronig residuals increase and the frequency points become less reliable. For frequencies above 100 kHz, the inductance is dominant and the cable calibration has a strong influence on the measured data. Therefore, the most reliable frequency range is chosen in order to judge the models' suitability.

First of all, the difference between transmission line and serial models is investigated, together with the influence of using an HN-element instead of an RQ-element. Therefore, Figure 7.9a shows the comparison of RMSE between model 1,2,3 and 4 for varying SOC. The influence of the applied impedance model on the fit result is mainly remarkable for low SOC which could already be seen for the fits previously presented. For high SOC, no model provides superior fit quality. From RMSE at low SOC, it can be concluded that the use of an HN-element instead of an RQ-element to describe the SEI (model 2) or using a TLM with fixed parameters (model 3) is preferable.

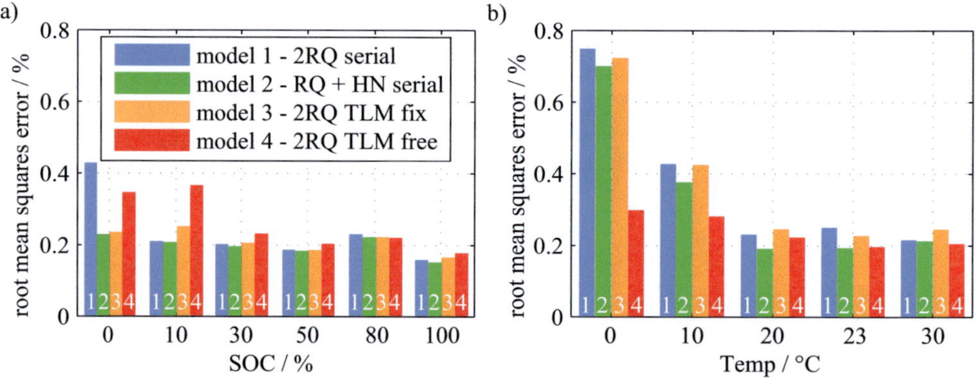

Figure 7.9: RMSE of fitting model 1-4 to the graphite-anode impedance spectrum for a) SOC-variation at 23 °C and b) Temperature variation at SOC=100 %.

Furthermore, it is important to analyze the fit quality at temperature variation, as that shows the changing influence of the TLM structure due to the temperature dependence of ionic conductivity. The corresponding residuals of model 1-4 are shown in Figure 7.9b. In this comparison, a clear separation concerning the fit quality can be observed at low temperatures as model 4 provides a superior fit quality. This demonstrates that the decreasing ionic conductivity at low temperatures increases the importance of the transmission line structure. The difference between model 3 and model 4 is caused by the different ionic resistance used in the TLM. Model 3 uses the theoretical ionic resistance which is calculated from electrolyte conductivity and microstructure parameters, whereas model 4 uses the ionic resistance fitted for SOC=100 % and 23 °C. The obtained resistance is a factor of six higher compared to the theoretically calculated values and therefore increases the impact of the TLM structure.

Another issue is the choice of a proper diffusion model. Three Generalized Finite Space Warburg elements were introduced in Ref. [54] and presented in this thesis in Section 2.4.1.4 in order to describe 1D- 2D- and 3D-diffusion. Figure 7.10a shows the RMSE for all three diffusion models which are implemented in model 1, 5 and 6. Again, the residuals increase for low SOC whereas they remain similar for almost all high SOCs. No significant difference can be found between the three diffusion models proposed regarding their ability to describe the impedance spectrum in the frequency range measured. According to the structure of graphite-anodes, consisting of graphene planes aligned in parallel (illustrated in Figure 2.5), a two dimensional diffusion process occurs. Therefore, and due to the comparable fit quality among the other diffusion models, the 2D diffusion model is applied for further analysis.

The last alternative model structure discussed in this section affects frequency range I. As already discussed, two possible loss processes might cause the polarization in this frequency range. The first interpretation for the high frequency loss process, assumed for all models previously compared, is the contact resistance between anode and current collector. Another interpretation would be the inter-particle resistance between graphite particles. This interpretation affects the position of the corresponding RQ-element inside the equivalent circuit model and leads to the previously introduced two path TLM. The RQ-element describing the high frequency loss process is therefore set into the second path of the TLM. Figure 7.10b compares the one path and two path TLM in order to check this alternative model structure. It is obvious for the entire SOC-range that the two path TLM is not able to describe the measured spectra. The residuals are larger than for the one path TLM throughout the entire SOC-range.

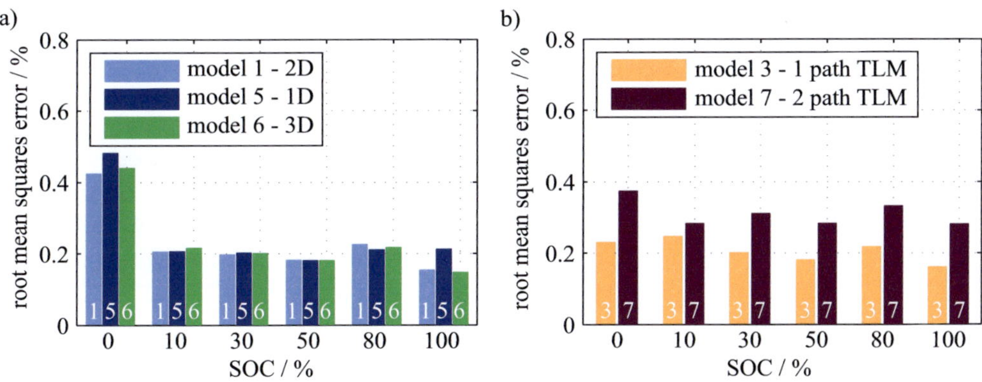

Figure 7.10: RMSE for a) different diffusion elements in model 1,5 and 6 and for b) two kinds of TLM structure in model 3 and 7, calculated from fitting the models to the graphite-anode impedance spectra for varying SOC at T=23 °C.

Relative residuals and RMSE are only a mathematical measure for fit quality concerning the accuracy when describing the measured impedance spectra. Model 4 is the best impedance model concerning this mathematical criterion for low temperatures. Models 2 and 3 are slightly better for low SOC. In the next step, the fitting results have to be checked on their validity with regard to their physical interpretation. Therefore, the following section introduces fit results of different impedance models and a recommendation concerning an optimal impedance model for the graphite-anode.

7.7 SOC-Dependency of Loss Processes

In this section, the fit parameters obtained for each loss process are compared in order to check the influence of the impedance models. The comparison covers ohmic resistance R_0, contact resistance $P_{CC,A}$, charge transfer process $P_{CT,A}$, SEI-process $P_{SEI,A}$ and solid state diffusion $P_{diff,A}$. As the two path TLM offered poor fit quality and 1D- and 3D-diffusion models did not have an effect on fit quality, only models 1 to 4 are taken into account for the following analysis. Figure 7.11 shows the comparison of SOC-dependent polarization for each loss processes.

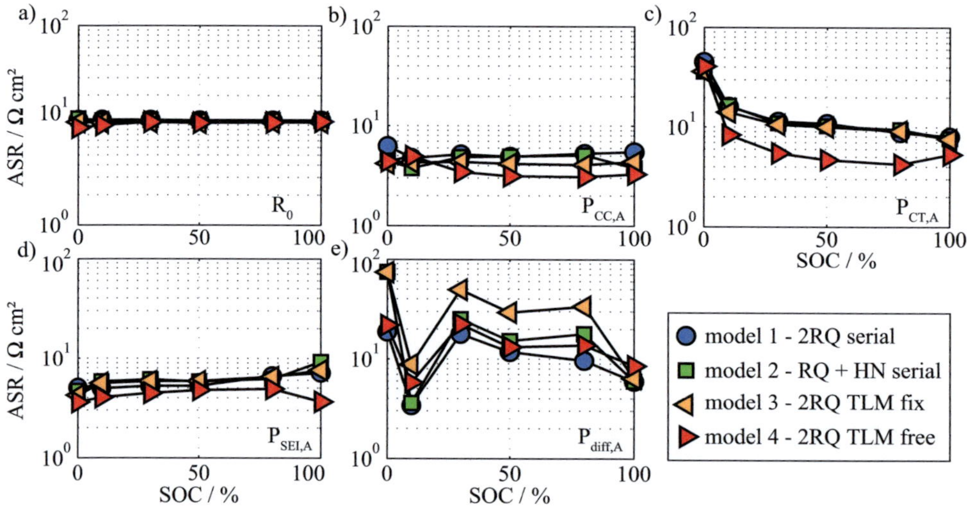

Figure 7.11: Polarization of anode loss processes for SOC-variation at T=23 °C, determined using model 1 to model 4. a) Ohmic resistance (R_0), b) Contact resistance anode/current collector ($P_{CC,A}$), c) Charge transfer anode/electrolyte ($P_{CT,A}$), d) SEI anode/electrolyte ($P_{SEI,A}$), and e) Solid state diffusion in the anode active material ($P_{diff,A}$).

The trend of SOC-dependency seen here is similar for all impedance models. All models reveal, as expected, an SOC-independence for the ohmic resistance. The contact resistance $P_{CC,A}$ shows only small variations for different SOC caused by the fit itself. However, no systematic SOC-dependence can be observed.

The ASR of contact resistance varies between $3\,\Omega\,\mathrm{cm}$ to $5\,\Omega\,\mathrm{cm}$ for model 4 and $5\,\Omega\,\mathrm{cm}$ to $6\,\Omega\,\mathrm{cm}$ for model 1. This emphasizes not only the importance of the correct impedance model choice, but also that the origin of these variations can be due to an interaction between different loss processes during the fit or, for $P_{CC,A}$, a different sensitivity on the inductance for higher frequencies. For charge transfer $P_{CT,A}$, a strongly increasing polarization can be identified for decreasing SOCs. This characteristic behavior fits very well with DRT evaluation and can be represented by all models. Model 4 delivers a lower polarization compared to the other models. This is caused by the TLM structure as a part of the entire polarization resistance is caused by the limited ionic conductivity inside the pores.

The ASR of SEI-process $P_{SEI,A}$ does not show a characteristic SOC-dependence. According to the DRT this is expected. However, small variations in polarization occur for each model and between the models. These variations are again caused by the fit stability as the interaction between charge transfer and SEI cannot be prevented due to their similar frequency region. Furthermore, the characteristic frequency for $P_{SEI,A}$ is not unambiguously fixed due to the broad peak distribution. The polarization of $P_{SEI,A}$ is again lower for model 4 as the electrolyte conductivity has a stronger contribution to polarization resistance compared to model 1-3.

A characteristic SOC-dependence can be observed for solid state diffusion for each impedance model. However, the strong variation in polarization between the impedance models and the unsteady trend of polarization indicate that the measured frequency range is not sufficient to investigate the solid state diffusion reliably. This recommends supplementing the EIS by additional time domain measurements as it will be presented for the graphite-anode in Chapter 8.

Figure 7.12 displays a comparison of SOC-dependency for the characteristic time constants of models 1-4. A similar trend can be observed as for the polarization. The time constant of contact resistance is very stable and almost the same for each model. The charge transfer time constant increases for decreasing SOC for all models and it is slightly higher for model 4. A significant difference between the models is obvious for the time constant of $P_{SEI,A}$. It is much higher for model 4 and shows a strong SOC-dependency. The higher time constant is caused by the strong impact of the TLM-structure for model 4 which is able to describe the decreasing peak sequence in frequency range II by the ionic resistance. Therefore, the time constant of $P_{SEI,A}$ moves close to the charge transfer in order to describe the second main peak P_2 correctly. This fact will be investigated in more detail for the temperature variation. The characteristic SOC-dependencies identified in this section are physically reasonable, as they describe the results of DRT-analysis adequately. However, a judgment of models 1-4 using these results is not possible as no fundamentally different behavior was discovered among the models. Therefore, the temperature dependence is investigated in the next section in order to get alternative criteria for model validation.

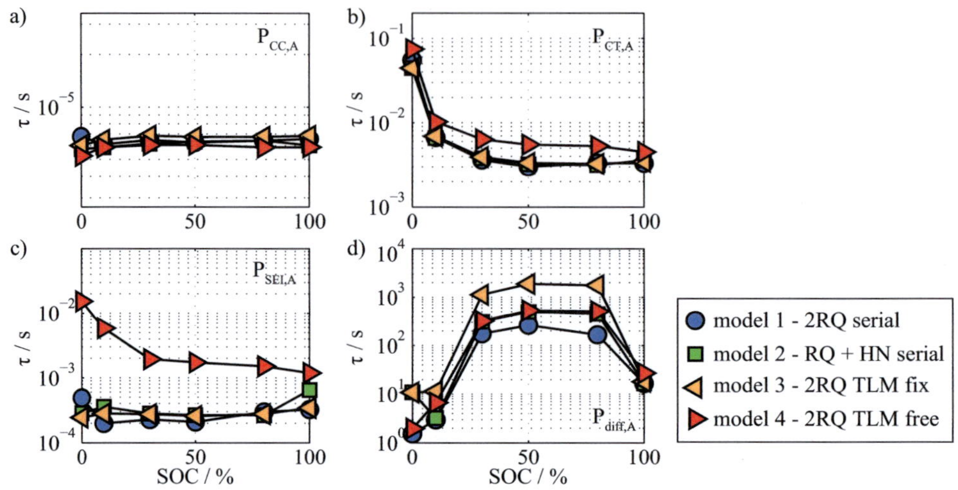

Figure 7.12: Time constant of anode loss processes for SOC-variation at 23 °C, determined using model 1 to model 4. a) Contact resistance anode/current collector ($P_{CC,A}$), b) Charge transfer anode/electrolyte ($P_{CT,A}$), c) SEI anode/electrolyte ($P_{SEI,A}$), and d) Solid state diffusion in the anode active material ($P_{diff,A}$).

7.8 Temperature-Dependency of Loss Processes

Figure 7.13 displays the temperature dependency of the polarization of each loss process. All loss processes are fitted with the Arrhenius-type Equation 2.7 in order to check the plausibility of the fit and to determine the activation energy of the underlying loss process. In Figure 7.13a, the ohmic resistance reveals a clear linear behavior in the Arrhenius plot for all models indicating that the ohmic resistance follows Arrhenius' law. This also holds true for $P_{CC,A}$ and $P_{CT,A}$, whereas $P_{CT,A}$ shows the strongest temperature dependence. Model 4 assigns the smallest polarization to these loss processes due to the additional contribution of the limited ionic conductivity inside the pores.

The most striking contrast between different models occurs for the polarization of $P_{SEI,A}$. Models 1-3 yield a similar polarization having a weak temperature dependence. In contrast, model 4 delivers a very strong temperature impact, leading to a smaller polarization for high and a larger one for low temperatures compared to the other models. This significant contrast necessitates a detailed analysis of the fit results in order to check their plausibility. Figure 7.14 shows the DRT calculated for the impedance measurement at 0 °C and SOC=50 % and the corresponding fit of model 3 and model 4.

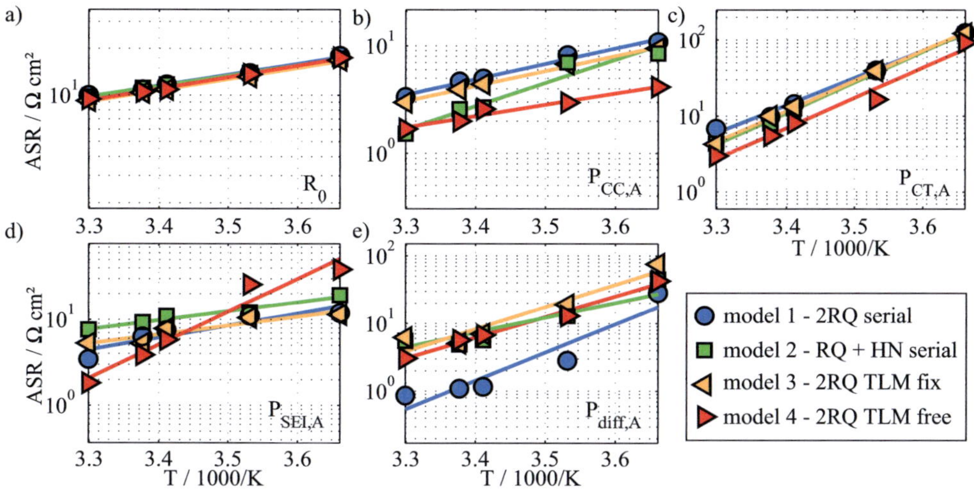

Figure 7.13: Polarization and Arrhenius fit of anode loss processes for varying temperatures at SOC=100 %, determined using model 1-4. a) Ohmic resistance (R_0), b) Contact resistance anode/current collector ($P_{CC,A}$), c) Charge transfer anode/electrolyte ($P_{CT,A}$), d) SEI anode/electrolyte ($P_{SEI,A}$), and e) Solid state diffusion in the anode active material ($P_{diff,A}$).

For model 3, the TLM parameters ionic conductivity and layer thickness were chosen according to the physical parameters previously calculated. This leads, as demonstrated in Figure 7.14a, to a small effect of the TLM-structure on the model impedance. For low temperatures, this yields that P_1 and P_2 are both described by the RQ-element for the charge transfer, whereas the peaks of the decreasing peak sequence are described by the RQ-element representing the SEI. As the temperature dependency of this peak sequence is very low, only a small temperature dependency for $P_{SEI,A}$ is identified. Assuming the previously described assignment of $P_{CT,A}$ and $P_{SEI,A}$ to the temperature-dependent main peaks P_1 and P_2, the fit result of model 3 leads to a deceptively low temperature dependency and an inaccurate activation energy. Model 4, being equivalent to model 3 but allowing a fit of the ionic resistance in the TLM, delivers other temperature dependencies which can be understood by analyzing Figure 7.14b. The RQ-elements, describing $P_{CT,A}$ and $P_{SEI,A}$, are fitted to the two main peaks in frequency range II. Furthermore, the effect of TLM-structure, originating from the limited ionic conductivity in the pores, is able to approach the characteristic TLM peak sequence for higher frequencies. This allows for a reliable determination of temperature dependencies which are supported by the DRT-analysis.

According to this analysis, the fit results and parameter dependencies of model 4 are more meaningful in terms of physical interpretation compared to the alternative model structures. In order to quantify the specific temperature dependency for each loss process, activation energies were calculated from the Arrhenius fits. Figure 7.15 shows

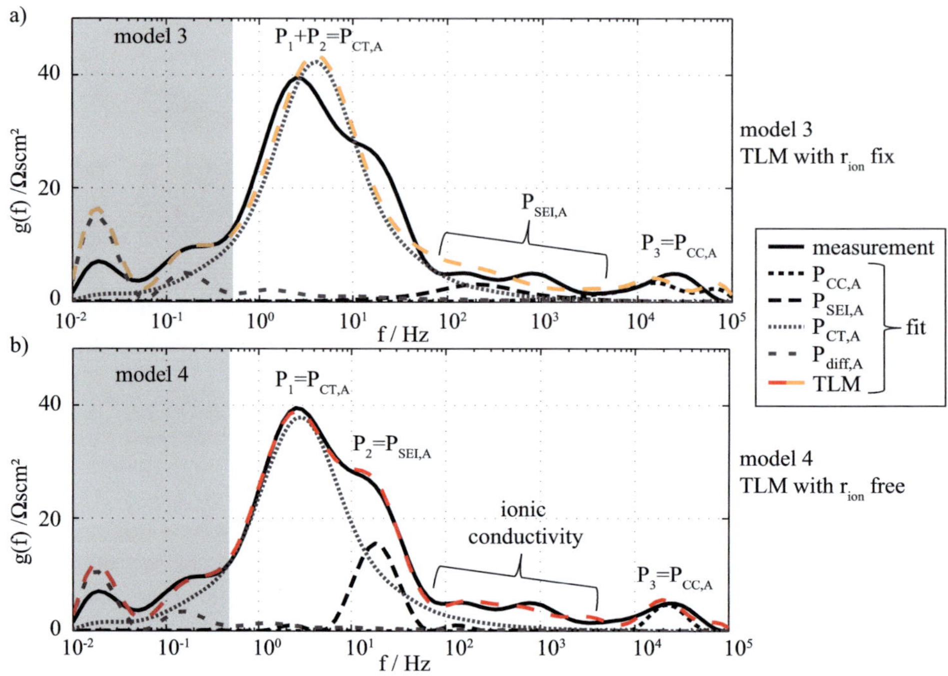

Figure 7.14: DRT of graphite-anode for 0 °C and SOC=50 % and DRT of the corresponding fit result for a) model 3 and b) model 4. The single impedance elements are plotted separately using the model parameters from fit in order to visualize the resulting position of impedance elements in the frequency range measured ($P_{CC,A}$: Contact resistance anode/current collector, $P_{CT,A}$: Charge transfer anode/electrolyte, $P_{SEI,A}$: SEI anode/electrolyte).

the activation energies for each loss process and for all models in order to quantify the possible effect of different impedance models. The ohmic resistance shows a similar activation energy (0.13 eV to 0.15 eV) for all models as it can be fitted in a very stable way. The activation energy of $P_{CC,A}$ varies from 0.21 eV for model 4 to 0.43 eV for model 2. This shows that not only $P_{SEI,A}$ depends on the choice of the correct impedance model structure in frequency range II, but also that the fit of frequency range I is affected by this choice as an interaction of impedance elements cannot be ruled out during the CNLS-fit. The charge transfer $P_{CT,A}$ is very dominant and is consequently identified clearly. The activation energy is therefore very similar around 0.79 eV for models 2-4 and it deviates slightly for model 1 (0.71 eV). The most significant difference can be observed for $P_{SEI,A}$ as already explained. Here, the activation energy for model 4 is determined to be 0.72 eV whereas the other impedance models deliver an activation energy between 0.2 eV and 0.3 eV. The activation energies are not identified for the solid state diffusion as the frequencies measured do not allow for a reliable evaluation.

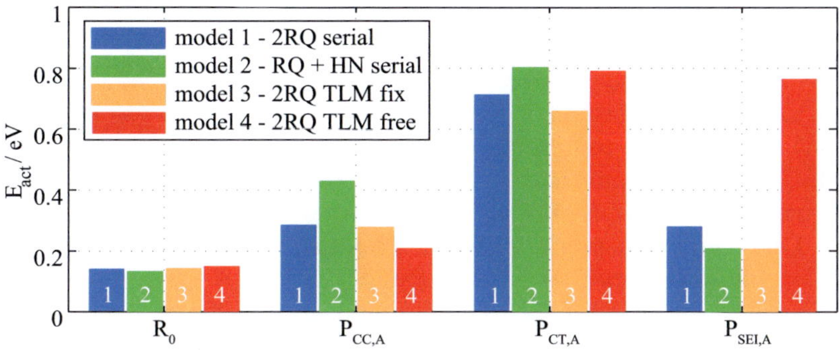

Figure 7.15: Activation energies for all loss processes determined by different models. More details about these models can be found in Table 7.1 and Figure 7.5 (R_0: Ohmic resistance, $P_{CC,A}$: Contact resistance anode/current collector, $P_{CT,A}$: Charge transfer anode/electrolyte, $P_{SEI,A}$: SEI anode/electrolyte).

7.9 Choice of Impedance Model and resulting Parameters

Three criteria are available to asses the proposed impedance models and to decide which impedance model is most suitable. The first criterion is the fit quality. The model for solid state diffusion did not show an effect on fit quality, wherefore 1D- and 3D-diffusion model were not investigated further. The 2D-diffusion model, which is physically most suitable, was applied in the subsequent analysis. The independence of fit quality on the applied solid state diffusion model shows that a measurement to lower frequencies is necessary in order to evaluate the solid state diffusion reliably. Moreover, the 2 path TLM in model 7 provided an insufficient fit quality compared to the other models, indicating that this model structure does not represent the physical processes. Therefore, model 1 to model 4 were used for the further analysis. Concerning the fit quality, model 4 delivered better fit results, especially for low temperatures, compared to models 1-3. The difference for other temperatures and SOCs is not remarkable. Especially for low SOCs and low temperatures, the fit quality gets worse for all models 1-4. As a consequence, the decision on the correct impedance model cannot be taken from the fit residuals.

The second criterion is the capability of the impedance models to represent the SOC-dependency of loss processes visualized by DRT analysis. The strong SOC-dependency identified for polarization and time constant of $P_{CT,A}$ by DRT-analysis could be represented by all impedance models. Also the weak SOC-dependency of polarization for $P_{CC,A}$ and $P_{SEI,A}$ can be represented properly for all models 1-4. In general, model 4 delivers smaller polarizations for $P_{CT,A}$ and $P_{SEI,A}$ compared to the other models due to the high contribution of the ionic conductivity in the TLM structure. The latter has naturally the maximum effect for model 4 as it assumes the lowest ionic conductivity. The last criterion applied in order to compare the impedance models was the identified temperature dependency. For this criterion, the activation energy of $P_{SEI,A}$ was most meaningful for a differentiation of the impedance models. It was shown by DRT analysis that model 4 best represents the temperature dependency of $P_{SEI,A}$ as the corresponding peaks are well fitted. The main reason for that is the strong effect of TLM structure in model 4, allowing for a description of the impedance spectrum measured. This leads to a much higher activation energy for $P_{SEI,A}$ compared to models 1-3, which is confirmed by DRT-analysis.

Therefore, model 4 is chosen for the following analysis in order to separate the overall losses in the cell. However, it is important to comment on the TLM parameters deviating from the expected values by using physical parameters. The layer thickness was fixed, for model 3 and model 4, to be $48\,\mu\mathrm{m}$ as it was determined unambiguously. The other relevant TLM-parameter is the ionic resistance r_{ion}. It was first calculated by ionic electrolyte conductivity and microstructure parameters as described in Section 7.5 and was subsequently used as fixed parameter for model 3 ($r_{ion} = 0.040\,\Omega\mu\mathrm{m}^{-1}$). It only changes for varying temperatures, coupled to the temperature dependence of the ohmic resistance. The ionic resistance was fitted at SOC$=100\,\%$ and T$=23\,^{\circ}$C for model 4, using the theoretically calculated value as starting parameter. By fitting this parameter, a 6.5 times larger ionic resistance ($r_{ion} = 0.262\,\Omega\mu\mathrm{m}^{-1}$) resulted for T$=23\,^{\circ}$C. There are several factors which may affect the ionic resistance and cause deviations from the expected values.

The microstructure reconstruction has several uncertainties concerning the parameters determined. The binder and a possible SEI layer cannot be measured. However, both are able to block the pores for lithium transport, leading to an overestimated porosity in the reconstructions. Furthermore, the anode is reconstructed in the delithiated state. Therefore, the active material expansion during intercalation causes a reducing porosity compared to the measured values. Another factor that can lead to an increasing ionic resistance is an incomplete filling of pores with electrolyte. This would decrease the active surface area and therefore increase the ionic resistance as well. Finally, it is not known whether an interaction between the lithium transport in the electrolyte and the active material surrounding the pores does exist. All these factors may increase the ionic resistance and impede an exact determination. A deviation by a factor of 6.5 is nevertheless very high. However, model 4 is used for the following analysis in spite of these limitations, as it allows the most meaningful fit in accordance with the DRT evaluation and the physical interpretation. Table 7.2 summarizes the identified loss processes, used equivalent circuit elements and determined parameters.

process	equivalent circuit element	f / Hz (SOC= 100 %, T=0 °C to 30 °C)	ASR / Ωcm^2 (SOC= 100 %, T=0 °C to 30 °C)	f / Hz (T=23 °C SOC=0 % to 100 %)	ASR / Ωcm^2 (T=23 °C, SOC=0 % to 100 %)	E_{act} / eV	physical origin
R_0	ohmic resistance		17.2... 9.2		9.4... 10.9	0.14	electrolyte (bulk)
R_{ion}	ohmic resistance (TLM)		52.5... 28.9		32	0.14	electrolyte (pores)
$P_{CC,A}$	RQ-element	21.7k... 42.9k	4.1... 1.7	30.1k... 27k	4.4... 3.3	0.21	contact resistance
$P_{SEI,A}$	RQ-element	13.9... 408.1	38.7... 1.8	10.9... 147.4	3.3... 3.4	0.72	SEI
$P_{CT,A}$	RQ-element	2.7... 48.7	93.4... 3.1	2.2... 35.4	41.4... 5.3	0.79	charge transfer
$P_{diff,A}$	2D-Warburg	0.7m... 9.2m	46.8... 3.0	85m... 5.9m	21.1... 8.5	-	solid state diffusion

Table 7.2: List of identified loss processes, their characteristic frequencies and polarization, evaluated applying model 4.

7.10 Transfer of Modelling Approach to other Graphite-Anodes

The impedance model introduced above was transferred to another graphite-anode in order to prove its general applicability and to check the influence of the TLM structure for different graphite-anodes. For this purpose, a high energy graphite-anode was chosen which was investigated by microstructure reconstruction in Ref. [1]. This additional anode will be denominated as anode B (Sanyo graphite-anode from Ref. [1] used in this section for comparison), whereas the anode investigated previously will be denominated as anode A (graphite-anode used in all other parts of this thesis) in the following analysis. There is a strong difference in both anodes' microstructure due to their different application. Anode A is extracted from a high power cell, optimized to achieve a high power density, whereas anode B is extracted from a high energy cell, optimized for high energy densities. The basic microstructure parameters of both graphite-anodes are compared in Table 7.3.

	Anode A (high power)	Anode B (high energy)
layer thickness	48 µm	76.1 µm
porosity ϵ	35.2 %	18.2 %
tortuosity τ	2.7	11.2

Table 7.3: List of microstructure parameters of high power (Anode A) and high energy (Anode B) graphite-anode.

The microstructure of both graphite-anodes differs strongly. First of all, the layer thickness is lower in the case of the high power anode. Furthermore, its porosity is higher, whereas its tortuosity is drastically smaller compared to the high energy anode. All of these differences indicate a minimization of lithium-ion transport losses through the electrolyte in the electrode pores in the case of the high power anode. This is a clear indication that the lithium-ion transport losses through the pores are a crucial factor for the electrode performance.

The parameters needed for the transmission line model introduced (model 4) which are specified by the electrode microstructure are given by the layer thickness L and the specific ionic resistance r_{ion}. Calculating the specific ionic resistance theoretically in accordance to Section 7.5 by using the microstructure parameters from Table 7.3 and the ionic conductivity of the electrolyte used in this cell (LiPF$_6$: σ_{ion} =1 S/m) leads to r_{ion} =0.242 Ωµm^{-1}. The latter is used as starting parameter to fit the impedance spectrum of anode B.

Figure 7.16 presents the impedance spectra, the DRT and the fit results of both graphite-anodes at SOC=100 % and T=23 °C. The impedance spectra of both anodes are basically similar, mainly consisting of a large flat semicircle for medium and a capacitive diffusion branch for low frequencies. The high frequency semicircle is more pronounced for anode A which can be an artifact of the electrode preparation. The DRTs of both anodes allow for a separation of charge transfer ($P_{CT,A}$), SEI ($P_{SEI,A}$), contact resistance ($P_{CC,A}$) and a decreasing peak sequence due to the limited ionic conductivity of the electrolyte in the pores.

Additionally, the fit results of model 4 and the simulation of all identified loss processes without the TLM structure using the previously determined fit parameters are depicted in Figure 7.16. The model is able to describe both impedance spectra very well. The effect of the TLM structure is visualized by the difference between the polarization of all single loss processes and the polarization obtained by integrating the latter into the TLM structure. It turns out that the effect of the TLM is drastically stronger for the high energy graphite-anode as the polarization of its interface loss processes are negligibly small compared to the overall impedance. This is an expected result as the significantly lower porosity and the drastically higher tortuosity of the high energy anode increase the resistance of the electrolyte in the pores responsible for the TLM behavior. Furthermore, the specific ionic resistance for anode B resulting from the fit is a factor of 2.15 higher than calculated theoretically. This effect has been observed more distinctive for anode A and its origin was discussed in the previous section.

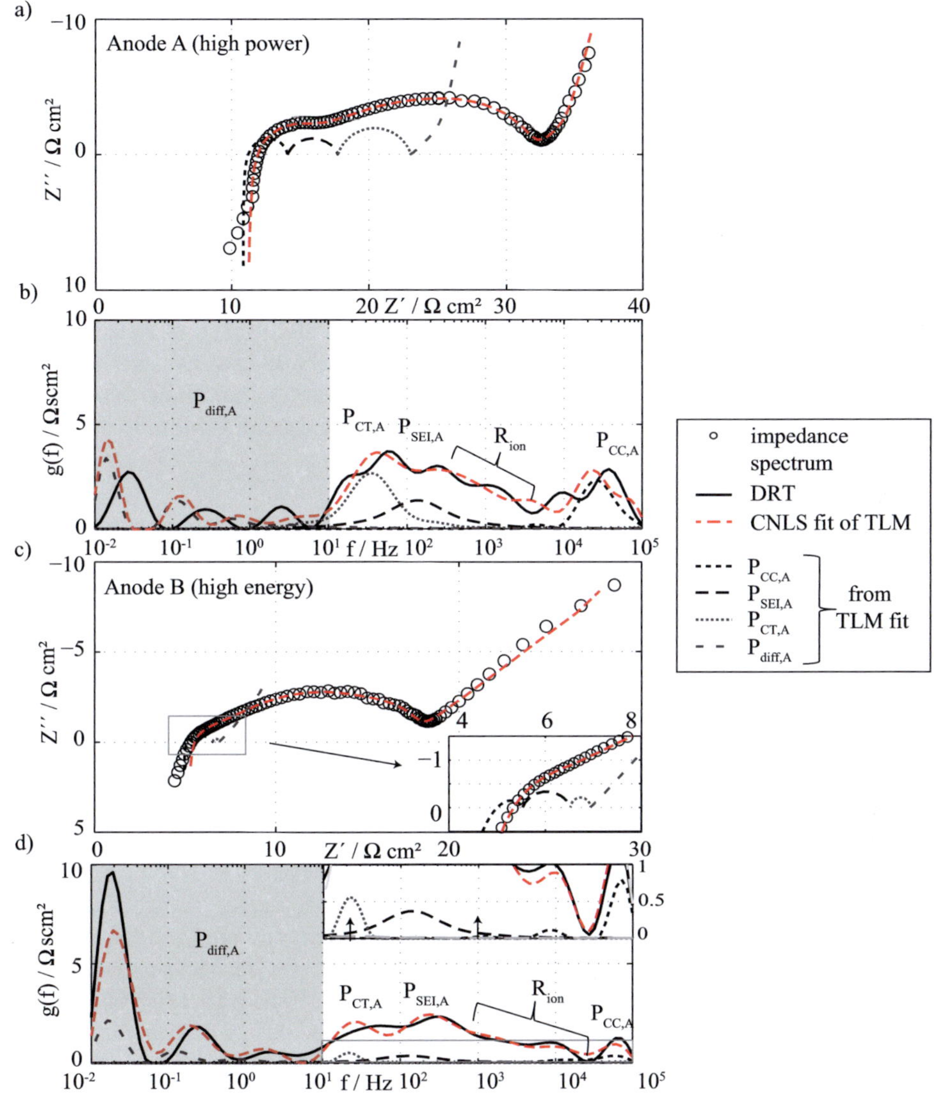

Figure 7.16: Nyquist plot and DRT of the impedance spectrum of a/b) high power anode (Anode A as investigated above) and c/d) high energy anode (Anode B, for comparison) for SOC=100 % and T=23 °C. Both measurements are approximated using model 4 introduced above, letting r_{ion} be a free parameter. The fit result of the entire TLM and the simulations of each single loss process using the resulting fit parameters are plotted (Model components: $P_{CC,A}$: Contact resistance anode/current collector, R_{ion}: Ionic resistance of the electrolyte in the pores, $P_{CT,A}$: Charge transfer anode/electrolyte, $P_{SEI,A}$: SEI anode/electrolyte, $P_{diff,A}$: Solid state diffusion in the anode).

The model is able to describe the DRT for both anodes very well. The frequency position of each loss process is in agreement with the previous interpretation and also the overall polarization and the TLM behavior are represented accurately. It is obvious just as for the impedance spectra that the major part of polarization is caused by the ionic conductivity and that this effect is stronger for the high energy anode.

This section demonstrated that the impedance model approach proposed in this chapter can be transferred to other graphite-anodes even if they exhibit a different microstructure. The impedance model elements just as the model structure are suitable to describe the occurring loss processes in agreement with the DRT evaluation. A comprehensive analysis of anode B would be possible in order to give a well-founded statement on the suitability of its physical interpretation. One of the main focus of such an analysis would be the question if the specific ionic resistance can be a fixed or whether it has to be a free parameter for this graphite-anode. However, there is no doubt that the model and its structure are suitable to describe the high energy graphite-anode as well.

7.11 Discussion and Conclusions

A comparison to published data is necessary for further discussion of the impedance model obtained and the interpretation proposed. Until today, a wealth of publications have dealt with graphite-like anodes and their loss processes. Depending on the active material used, the cell setup and the conducted measurements, a different number and physical interpretation of loss processes has been published. Usually, two [43,44,59,111,112,115,116,165,166] or three [114,167] loss processes are identified for graphite-like anodes (in addition to solid state diffusion). Several physical interpretations are provided, ranging from one or two charge transfer processes, SEI, graphite/current collector contact resistance to inter-particle contact resistance. Now, the physical interpretation here proposed is compared to the literature data.

The most dominant contribution is uniformly assigned to the charge transfer process between graphite and electrolyte [43,44,111,112,165,168]. Its position in the impedance spectrum is usually in a frequency range from 0.1 Hz to 10 Hz [112,115,165] between a high frequency semicircle and the characteristic bend to the diffusion branch. The strong SOC-dependency occurring for low lithium insertion levels found in this study can be confirmed by several experiments published. Chang et al. [112] found an increasing polarization above 0.3 V, whereas no SOC-dependency was found for lower potentials. Abe et al. [168] and Yamada et al. [165] observed a strong SOC dependency above 0.2 V. Furthermore, Umeda et al. [114] identified two charge transfer processes both showing an increasing polarization from 0 V to 0.3 V. The identified charge transfer processes are in the same frequency range and show a similar SOC-dependency as in this study, suggesting an agreement of the gained interpretation.

Some of these publications also analyze the temperature dependence of the charge transfer process. In Refs. [43,44], the charge transfer was identified as main contribution. In order to understand the underlying physics in more detail, the electrolyte components were varied and their influence on activation energy was checked. This experiment discovered a correlation between EC/lithium-relation and the activation energy of charge transfer, leading to the conclusion that the rate determining step for this charge transfer process depends on the desolvation of lithium-ions. These results were also supported by investigations of Abe et al. [168, 169] who did similar experiments. Applying the same solvent as in this study ($LiPF_6$ in EC:EMC, 1 : 1) leads to an activation energy of 0.73 eV [44] which is close to the 0.79 eV determined here.

In contrast to all these studies, an identification of a second process causing the main semicircle is inevitable in this study as the DRT undoubtedly shows two loss processes. This additional contribution is assigned to the SEI, as it shows weak SOC-, but a strong temperature dependency of 0.72 eV. It contributes significantly to the polarization for SOCs above 10 %.

In this study, another semicircle was identified for frequencies from 20 kHz to 40 kHz and assigned to the contact resistance anode/current collector. This additional high-frequency loss process is also identified in most published studies [43, 102, 111, 112, 114–116, 165]. Concerning its physical origin, there are different interpretations. The most popular interpretation for the high frequency arc is the lithium transport through the SEI layer [43, 44, 59, 114, 116, 165, 166], although a partial contribution of current collector/graphite interface is not excluded [102]. Holzapfel et al. [115] do not support the SEI as physical origin for the high frequency loss process as it already occurs previously to the SEI formation. Therefore, contact problems inside the electrode as grain-grain or graphite/current collector are supposed to play the major role for high frequency losses. Chang et al. [112] identified the contact between graphite and current collector as physical origin as it depends on electrode area and not on active surface area. The multitude of number and interpretation of high frequency loss processes indicates that the influence of contact resistance to current collector, inter particle resistance and transport through the SEI are strongly dependent on the active material, the fabrication of the electrode layer and the current collector. Depending on the electrode, some of these losses do not occur in general or dominate the high frequency response. Furthermore, it is possible that they have similar time constants and therefore overlap in the impedance spectrum.

A good example for overlapping loss processes is given in Ref. [59]. Herein, a broad semicircle is identified and described by four RC-elements with time constants reaching from 217 kHz to 150 Hz. Several SEI-layers are proposed as physical origin for these contributions. However, an unambiguous separation of the latter is not possible without using the DRT analysis. The approach of using several RC-elements increases the fit accuracy at the expense of decreasing fit stability.

A physically more meaningful model was presented in Ref. [102]. There, a TLM is applied in order to include the microstructure into the equivalent circuit model. A combined fit of several impedance spectra with equal TLM parameters is used in order to obtain stable and physically meaningful fit results. The fit results indicate a large contribution of the graphite layer which is assigned to the electronic layer conductivity. However, the model structure applied does not allow for a distinction between the contributions of the electronic path and the contributions of the ionic path. A meaningful assignment to one of them is not possible because the real microstructure parameters and the electronic layer conductivity are not known.

In this thesis, the ionic conductivity inside the pores was identified as a main reason for the depressed semicircle in the impedance spectrum of graphite-anodes. Applying the DRT method, microstructure reconstruction and TLM allowed for a quantification of the ionic resistance inside the pores. Depending on the temperature, this causes a main part of the polarization. Figure 7.16a demonstrated the impact of ionic conductivity on the impedance spectrum for T=23 °C and SOC= 100 %. The TLM is able to describe the impedance spectrum well, whereas the single loss processes as $P_{CC,A}$, $P_{SEI,A}$ and $P_{CT,A}$ cause only a part of the polarization.

In this chapter, the investigation of complex and – compared to lab-scale – thick anodes as they are used in advanced lithium-ion batteries has been presented. The identification of loss mechanisms is impeded by an interaction of the electrode microstructure and the relevant electrochemical loss mechanisms. A separation of these effects can be achieved by applying the DRT and supplementing this analysis with microstructure parameters. This procedure leads to the conclusion that more complex impedance models such as TLM are needed in order to describe these impedance spectra properly. This is important for any kind of impedance analysis in which physically relevant parameters are to be determined. Not only the identification of aging mechanisms is affected, also the physically motivated time domain simulation has to take this correlation into account. The interaction was shown to be even stronger for electrodes from high energy density cells as the latter provide a larger electrode thickness and tortuosity and a lower porosity compared to high power electrodes.

8 Multi-Step Approach for the Analysis of 18650 Lithium-Ion Cells

This chapter introduces the analysis of a commercially available high power 18650 cell applying a combination of EIS and TDM measurements and experimental cells. The identification of internal loss mechanisms in commercial cells is a crucial step for a targeted improvement of today's lithium-ion cell generation. Furthermore, an understanding of degradation mechanisms is one of the main goals of current research activities in the field of lithium-ion batteries. This understanding is crucial for a lifetime prediction and an assessment of the liability risk at an extensive use of high performance lithium-ion cells in electric vehicles.

The internal resistance of lithium-ion batteries can be used as a measure for their performance or state of health. It is usually determined by current step response or electrochemical impedance spectroscopy [53, 75, 76]. Hereby, electrochemical impedance spectroscopy is the more powerful method as it allows for the separation of individual loss processes. However, the application of electrochemical impedance spectroscopy on commercially available or automotive cells comes along with several disadvantages regarding the investigation of the frequency range in which electrochemical loss processes occur. Figure 8.1 shows an impedance spectrum of a commercially available 18650 cell. One limitation for the analysis of commercial cells appears for high frequencies. The impedance response of the investigated cell is dominated by an inductance for frequencies above 5 kHz (see Figure 8.1). This inductive behavior is caused by the cables of the measurement equipment as well as by the current collectors of the lithium-ion cell itself. A smaller impedance increases the influence thereof. This is why the evaluable frequency range differs for different cell types according to their polarization and allows therefore the analysis of a cell dependent and limited frequency range.

Another challenge for commercial cells is the separation of anode and cathode losses. Figure 8.1 shows just one characteristic semicircle dominating the impedance spectrum. Considering the variety of eventual loss processes which might occur at each of the electrodes, this seems to be a combination of several overlapping loss processes. The latter are obviously not separable by applying standard EIS on the full cell.

For low frequencies, the impedance response is dominated by the intercalation capacity. This problem can be solved by a pre-processing step as introduced in Chapter 5. However, this procedure does not enable the analysis of lower frequencies. The measurement of frequencies below 10 mHz by standard EIS is not practical as this increases the measurement time drastically.

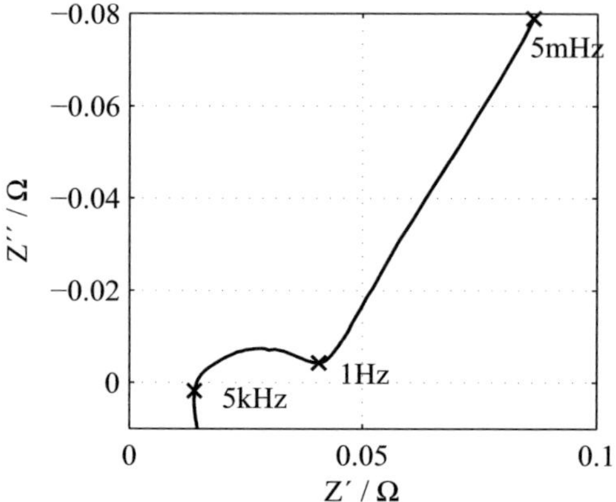

Figure 8.1: Impedance spectrum of a commercially available 18650 lithium-ion cell.

The major focus of this chapter is on three targets:

1. Increase of the evaluable frequency range to higher and lower frequencies.
2. Superior separation of overlapping loss processes.
3. Physical interpretation of the 18650 cell impedance spectrum.

The following approach is applied in order to fulfill these requirements. The impedance response of a commercial 18650 cell is characterized by combining electrochemical impedance spectroscopy and time domain measurements. This provides information over a wide frequency range of 11 decades from $1.5\,\mu Hz$ to $100\,kHz$. Subsequently, the impedance spectra are evaluated by the DRT in order to analyze the relevant frequency ranges separately with a high resolution. Furthermore, the occurring processes are assigned to anode, cathode, or electrolyte, or interfaces thereof. For this purpose, the 18650 cell is first measured, then opened in a glovebox and the extracted electrodes are re-analyzed in experimental cell setups. The contributions of cathode and anode are separated using a reference electrode. Furthermore, the evaluable frequency range is expanded, as impedance spectra of experimental cells are less affected by inductive contributions compared to 18650 cells. This allows for a comparative study of measured impedance spectra and the calculated continuous DRT of experimental cells with the original 18650 cell. New insights into the internal loss processes can be given as the DRT visualization directly provides information about the characteristic frequencies of loss processes. This enables an independent evaluation of various frequency ranges and a detailed comparison of experimental and commercial cell measurements via polarization and characteristic frequency of the occurring loss processes.
Most of the results presented in this chapter have been published in Ref. [60].

Figure 8.2: *left:* SEM-image of the investigated LiFePO$_4$-cathode and *right:* graphite-anode.

8.1 Investigated 18650 Cell

In this study, Sony 18650 high power cells (*SE US*18650*FT*, basis data in Appendix A) were analyzed and opened afterwards. According to the data sheet, these cells have a nominal capacity of 1.1 Ah for a 1C discharge between 2.0 and 3.6 V. The cells were opened in a discharged state as described in Section 4.3 and afterwards the electrodes were analyzed in experimental cell housings. The cell contains a LiFePO$_4$-cathode (d=56 µm) and a graphite-anode (d=48 µm). Figure 8.2 shows the cathode consisting of a very complex microstructure which contains small primary particles (average particle diameter 218 nm) and larger active material agglomerates (average agglomerate diameter 1.25 µm). Furthermore, vapor grown carbon fibers (VGCF) are added to the composite electrode in order to increase its electronic conductivity and binder is added in order to fix the electrode composite. The graphite-anode contains relatively large graphite particles (average particle diameter 4.236 µm) and binder.

The electrodes were analyzed at IWE by Moses Ender [147, 159] using FIB/SEM for LiFePO$_4$-cathodes and micro X-ray tomography [1, 164] for graphite-anodes in order to determine their microstructure parameters as porosity, particle size and material fractions. Table 8.1 shows these values for both electrodes.

One noticeable fact is the different particle size for cathode and anode particles which is already obvious in the SEM-pictures. The average anode particle size is almost 20 times larger than the cathode particle size. This results in a far smaller specific active surface area of 0.395 µm^{-1} compared to 6.301 µm^{-1}. The smaller active surface area limits the lithium intercalation into the graphite. However, this is a compromise between limited lithium intercalation for large particles and high irreversible loss of active lithium by excessive SEI formation for small particles. The porosity is slightly larger for the cathode microstructure.

	LiFePO$_4$-cathode	graphite-anode
volume fractions		
active material	0.552	0.648
carbon black additive	0.064	-
pores	0.384	0.352
average particle size / µm		
active material particles	0.218	4.236
active material agglomerates	1.25	-
specific surface area / µm^2/ µm^3		
active material	6.771	0.395
carbon black additive	2.037	-
pores	7.867	0.395
active surface area	6.301	0.395

Table 8.1: Overview of microstructure parameters, determined by Moses Ender [1].

8.2 Experimental Cell Measurements

The next step after opening the 18650 cell is the investigation of its electrodes in experimental cells. This increases the high frequency limit from 5 kHz to 100 kHz due to a less pronounced inductance for experimental cells. Furthermore, the experimental cells can include a reference electrode which allows for the separation of anode and cathode losses. In this section, the preparation of the extracted electrodes, the investigation of these electrodes by EIS and TDM and the analysis of its polarization losses by impedance modelling are discussed.

8.2.1 Experimental Cell Configurations and Reproducibility

Cathode and anode half-cells were assembled and measured in order to investigate reproducibility and capacity of the harvested electrodes. The main emphasis of half-cell investigations was put on capacity and for the cathodes on impedance, which is influenced by the preparation of double side coated cathodes. All half-cells were assembled in EL-Cell housings [135] with LiFePO$_4$-cathode or graphite-anode as working electrode and lithium metal as counter electrode. A 1.55 mm thick glass fiber separator of EL-Cell was used as separator and electrolyte reservoir. The electrolyte was a standard electrolyte (*LP*50 from BASF) based on an EC:EMC (1:1) solvent mixture with 1 M LiPF$_6$ as conducting salt.

Full cells with and without steel mesh as reference electrode were used in order to identify and separate the electrochemical loss processes by EIS and TDM. Four 220 µm thin glass fiber separators were utilized as spacer in order to separate anode and cathode from the reference electrode mesh. A 1 M mixture of LiPF$_6$:LiClO$_4$ (1:9) in

cell configuration	full cell	LiFePO$_4$ half-cell	graphite half-cell
upper cut-off voltage / V	3.6	3.7	0.5
lower cut-off voltage / V	2.0	2.6	0.01
current / mAh	3.5	3.5	0.7
CV-phase during	charge	charge	discharge
cut-off current / mAh	0.35	0.35	-0.07

Table 8.2: Settings for the cycling of measured cell configurations.

EC:EMC (1:1) solvent mixture was used as electrolyte in the full cells which were assembled to conduct impedance analysis over the entire parameter field. This mixture is reported to prevent Fe-dissolution from LiFePO$_4$ and to passivate the aluminum current collector [170, 171]. This increased the stability of the electrodes' impedance for the extended measurement time. 5 % of vinylene carbonate (VC) was added to the electrolyte for those cells assembled for time domain measurements in order to further increase their stability [46, 47].

The calculation of the area specific capacity which is derived from 18650 cell capacity (1.1 Ah) and electrode area (759 cm^2) reveals 1.45 mAh/cm^2. This capacity, which corresponds to 3.68 mAh per EL-Cell, is expected to be provided by both electrodes at least. Depending on the matching of both electrodes in the full cell, the graphite-anodes deliver higher capacities as the anode is usually larger than the cathode capacity. The current which is adjusted for LiFePO$_4$ half-cells and experimental full cells is almost 1C (3.5 mA), whereas the graphite half-cells were cycled with a fifth of this current (0.7 mAh). This is due to the low potential of graphite in the lithiated state. The overpotential at high currents therefore leads to an incomplete discharge process of graphite half-cells. Table 8.2 lists currents and the charge/discharge voltage for each cell configuration.

Single side coated graphite-anodes were extracted directly from the outer layer of the 18650 cell. Before the measurement in experimental cells, they were punched out and washed in EMC. Afterwards, they were dried for ten minutes in a vacuum and assembled in half-cells. Figure 8.3a presents the mean value and standard deviation for the formation cycles of graphite half-cells. The charging capacity of graphite half-cells is shown as this corresponds to the discharge of the full cells. The graphite half-cells provide a very stable average capacity of more than 2.1 mAh/cm^2, which is about 45 % more than the capacity expected. The capacity does not decrease due to SEI formation as the lithium reservoir in the lithium metal anode is much larger than the capacity needed.

As the LiFePO$_4$/aluminum sheets are always coated on both sides with cathode material, one coating has to be removed by a dissolving liquid and a cleaning medium. Various solvents were tested to remove the cathode layer. The first solvent applied was NMP, as NMP itself is used as standard solvent to mix a cathode slurry. It is able to dissolve the standard cathode binder Polyvinylidenfluorid (PVDF) and is therefore able to dissolve

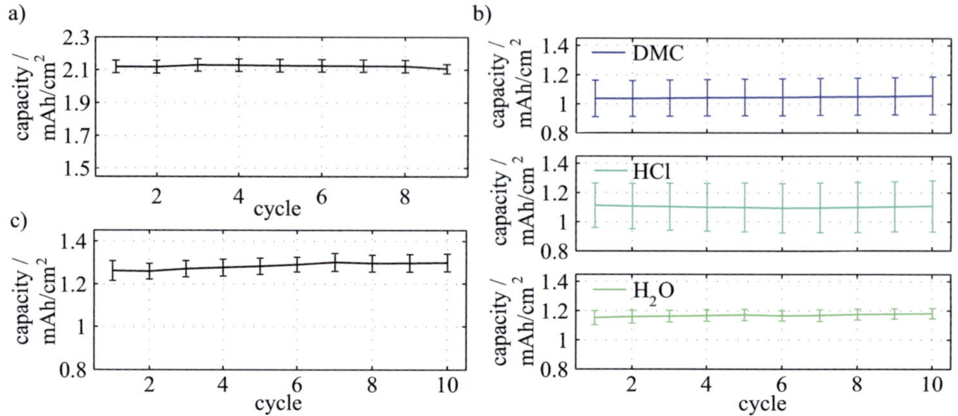

Figure 8.3: Mean value and standard deviation for a) the first charge cycles of graphite half-cells, b) the first discharge cycles of LiFePO$_4$ half-cells with differently prepared cathodes and c) the first discharge cycles of experimental full cells.

the cathode slurry. The tests yielded that the use of NMP leads to strong smearing of the cathode layer and does not remove it thoroughly. Furthermore, it is not known whether the remaining NMP can be removed properly. Therefore, alternative solvents were tested.

Water (H$_2$O) and diluted hydrochloric acid (HCl) were tested as they were expected to dissolve the PVDF itself. The use of water and water containing solvent necessitates an outsourcing of the preparation procedure from the glovebox into a flow box as the latter one is not as sensitive to increasing water contents. The last solvent applied was dimethyl carbonate (DMC), which acts as a transport medium to remove the cathode layer. For all preparation methods it is important to protect the back side of the cathode from the solvent in order to prevent an influence on the cathode layer investigated later. Furthermore, the mechanical stress should be kept to a minimum. The preparation itself delivered water and diluted HCl as good solvents, as they allow a large-scale removal of cathode layers without high mechanical efforts. The critical point at this preparation method is the use of water, as water is very critical for the cells' stability. Furthermore, the flow box itself has an increased water content which can also decrease the cathode performance.

For all preparation methods it is necessary to wash the electrodes after the preparation in order to remove the remaining solvent or electrode material and to wash the remaining electrolyte from the electrode surface. This washing process is done by EMC as it is also a component of the electrolyte used subsequently. Afterwards, DMC prepared electrodes are dried for ten minutes in a vacuum. Using a water containing solvent necessitates an extensive drying for at least 3 h at 100 °C in a vacuum in order to remove the entire dissolving liquid. Having been prepared this way, anodes and cathodes are stored under glovebox atmosphere and assembled in experimental cells. The results of experimental cell measurements with electrodes differently prepared are discussed in the following.

Figure 8.3b shows the mean value and the standard deviation for the formation cycles of $LiFePO_4$ half-cells for different preparation methods. The values for DMC-prepared, HCl-prepared and H_2O-prepared $LiFePO_4$-cathodes are presented in Figure 8.3b. All average capacities are below the expected capacity, which is set as maximum value of the Y-axis. The highest average capacity and the smallest variation is obtained by the H_2O-prepared electrodes ($C_{avg}(10)$=1.19 mAh/cm^2 and $C_{std}(10)$=2.9 %). This yields H_2O as best preparation method.

However, there are several conceivable reasons why the capacity of all cathodes prepared is below the expected capacity:

1. The preparation of electrodes can lead to inactive material fractions, which are disconnected from the remaining composite electrode. This can explain the different influence of preparation methods on capacity. The H_2O-preparation is therefore the most promising as the average capacities are higher compared to the other preparation methods.

2. A passivation layer consisting of electrolyte components, which can not be washed off completely might cover the cathode and is resolved by the new electrolyte during formation. This would explain a part of the smaller capacity as well as the increasing capacity during the formation cycles as the removed passivation layer can release access to additional active material.

3. Another possible reason for a decreased capacity can be found in the complex microstructure of the $LiFePO_4$-cathode. As shown in Figure 8.2, there are two kinds of agglomerates of active material. One part is very dense which means that almost no electrolyte can penetrate the agglomerates. The other part of agglomerates has a certain, but very small porosity. The electrolyte can find its way into these open pores, however this is a very slow process which might explain the slightly increasing capacities during the first cycles.

4. Lastly, the discharge cut-off voltage is chosen to 2.7 V for $LiFePO_4$ half-cells which is significantly higher than the full cells' discharge voltage (2.0 V). This was done in order to use a safe voltage range as the final discharge potential of the $LiFePO_4$-cathode in the full cell was not known.

An identification of the main reason for a reduced capacity is not possible at that time. However, the following section about experimental full cells further elucidates the capacity losses observed.

Figure 8.3c depicts the mean value and the standard deviation for the formation cycles of experimental full cells using H_2O-prepared cathodes. The capacity increases during the first seven cycles and converges towards 1.29 mAh/cm^2 for the tenth cycle. This final capacity is slightly higher than the final $LiFePO_4$ half-cell capacity. However, it is 11 % bellow the expected capacity.

The coulombic efficiency of an experimental full cell is shown in Figure 8.4b in order to explain its capacity behavior. For the first cycle, the coulombic efficiency reaches 92.5 %, which means that 7.5 % of the active lithium is consumed for SEI formation

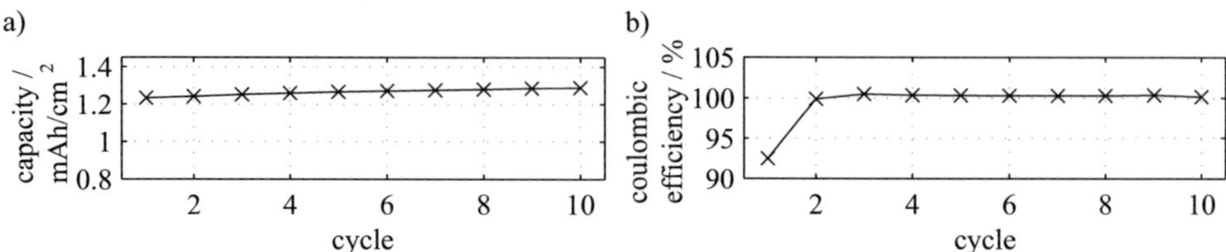

Figure 8.4: Capacity and coulombic efficiency of first discharge cycles of one experimental full cell.

during the first cycle. The remaining 3.5% of capacity lack must be caused by one of the effects (one, two or three) described for the $LiFePO_4$ half-cells. From the third cycle, the coulombic efficiency is slightly above 100% which leads to an increasing discharge capacity. This can be explained by a gain of active material and lithium during the first cycles caused by the second and third effect mentioned above.

These results from full cell analysis demonstrate that the main part of capacity loss in $LiFePO_4$ half-cells must be caused by the different discharge cut-off voltage. Therefore, the high average capacity and the low standard deviation identifies H_2O as the best cathode preparation technique. The standard deviation of $LiFePO_4$ half-cell as well as full cell capacities shows a sufficiently high reproducibility of experimental cells concerning these parameters. Another important factor is the reproducibility of the impedance spectrum of extracted $LiFePO_4$-electrodes.

Impedance spectra were measured for all $LiFePO_4$ half-cells assembled in order to investigate the different preparation techniques. The half-cell impedance model which was presented in Section 6.3 was used in order to analyze the loss processes and their variation separately as it is not meaningful to compare the impedance spectra directly with the naked eye. The applied model consists of inductance, ohmic resistance, high frequency cathode loss process (P_{2C}), lithium loss process (P_{1A}) and solid state diffusion ($P_{diff,C}$), their physical interpretation was already introduced in Chapter 5. The cathode charge transfer P_{1C} is hereby neglected as it cannot be identified in the half-cell impedance because P_{1C} is much smaller than P_{1A} and overlapped by the latter.

Figure 8.5 compares the variation of all relevant model parameters for different preparation methods. The first parameter, important to judge the reproducibility of cell assembly, is the ohmic resistance. It is dominated by the electrolyte resistance as the separator and therefore the electrolyte in experimental cells is usually very thick compared to 18650 cells. The average value is between $21.9\,\Omega cm^2$ for DMC- and $25\,\Omega cm^2$ for H_2O-preparation. The higher ohmic resistance of H_2O- and HCl-prepared electrodes can be a result of a passivation layer on the back side of the current collector caused by H_2O- and HCl-preparation which increases the resistance between current collector and cell housing.

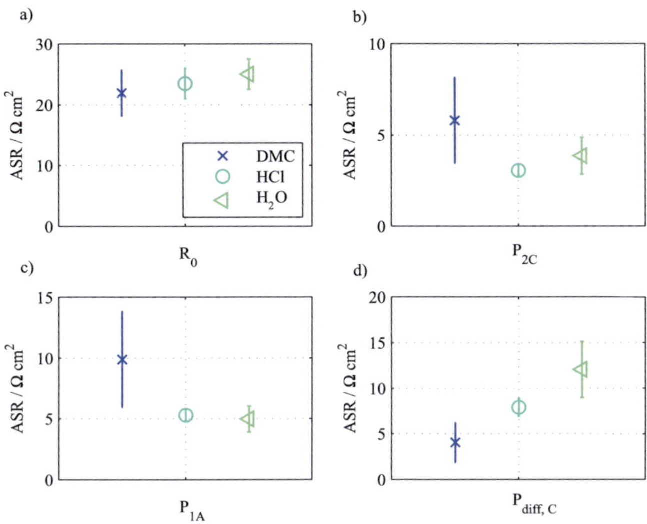

Figure 8.5: Variation of model parameters: a) ohmic resistance R_0, b) contact resistance P_{2C}, c) lithium loss process P_{1A} and d) solid state diffusion $P_{diff,C}$ in the cathode active material.

The second parameter, indicating a good cathode preparation, is the contact resistance between cathode and current collector P_{2C} (see Figure 8.5b). This value is assumed to depend on the mechanical stress caused by the electrode preparation. The average value for P_{2C} depends on the preparation method. It varies from $5.8\,\Omega\text{cm}^2$ for DMC over $3\,\Omega\text{cm}^2$ for HCl to $3.8\,\Omega\text{cm}^2$ for H_2O as preparation method. This variation shows that the preparation itself has an influence on contact resistance. Every preparation method has a certain effect on the contact resistance, and this must be taken into account for the following 18650 cell analysis. The variation is only $9.5\,\%$ for HCl-preparation whereas it reaches $26\,\%$ for H_2O-preparation. Concerning the average contact resistance P_{2C} and its variation, HCl delivers the best results followed by H_2O. DMC delivers the worst reproducibility.

The variation of P_{1A} usually corresponds to the variation of cell capacity as the lithium anode impedance depends on the amount of cycled lithium (see Section 5.2.3). This holds true for the DMC- and H_2O-prepared electrodes in Figure 8.5c. For the HCl-prepared electrodes it is different as the impedance spectra were measured not always at $100\,\%$ SOC. The ASR of solid state diffusion is illustrated in Figure 8.5d. It also depends on the state of charge of each cell. Even the cells measured at $100\,\%$ SOC show a variation in their open circuit potential. This influences the intercalation capacity strongly and has therefore an effect on the solid state diffusion. However, this variation is not meaningful for the validation of the electrode preparation.

In summary, using H_2O is the best preparation method concerning its specific capacity, followed by DMC and HCl. The analysis of the loss processes by impedance spectroscopy reveals HCl as best preparation method followed by H_2O and DMC. Therefore, H_2O is chosen because it seems to be the best compromise between a superior capacity and an acceptable variation of impedance spectra. All following measurements are conducted with H_2O-prepared electrodes.

8.2.2 Identification of High Frequency Losses via EIS

All results presented in this section were obtained from experimental full cells with reference electrode, assembled like introduced above. After assembly, the full cells were cycled between 2.0 and 3.6 V for twenty times with a current of 3.5 mA obeying a CCCV protocol. Then, impedance spectra were measured under SOC variation (100 %, 80 %, 50 %, 30 %, 10 %, 0 %). The SOC was stepwise decreased from 100 % to 0 %, whereas it was defined by the integration of discharge current and the previously measured discharge capacity. Next, impedance spectra were measured for varying temperatures (30 °C, 23 °C, 20 °C, 10 °C, 0 °C) at three different SOCs (100 %, 50 %, 0 %). All measurements were conducted using the working electrodes for the excitation signal and measuring the voltage between working electrodes and between working and reference electrode simultaneously. Therefore, three impedance spectra result for each impedance measurement: (i) The full cell, (ii) the cathode and (iii) the anode impedance spectrum. All impedance measurements were performed in a frequency range from 1 MHz to 10 mHz with an stimulation of 10 mV rms for high and 5 mV rms for low frequencies (f<100 Hz).

8.2.2.1 Separation of Anode and Cathode Losses

Figure 8.6 presents a typical impedance spectrum of an experimental full cell with reference electrode. Full cell and half-cell impedance spectra are plotted separately. The full cell impedance spectrum reveals the capacitive diffusion branch for low frequencies and the inductive behavior for frequencies above 100 kHz. In addition, two characteristic semicircles occur. Regarding the impedance curves of anode and cathode, two semicircles can be identified for the anodic and one semicircle for the cathodic electrode. The ohmic resistance splits up equally for both electrodes, which indicates an optimal position of the reference electrode. Further, no inductive loops or other typical artifacts can be observed for the impedance measurements via reference electrode [134, 141].

Figure 8.7 displays the corresponding DRTs in order to give a better separation of anode and cathode contributions. The measured frequencies are divided into three characteristic frequency ranges, denominated as I, II, and III. Frequency range I includes the high frequencies between 5 kHz and 100 kHz. These frequencies are usually part of the ohmic resistance for commercial cells (see Figure 8.1) as the inductance dominates for frequencies above 5 kHz. For experimental cells, this frequency range can be analyzed by DRT analysis. Here, two characteristic peaks are clearly distinguishable in the calculated DRT of the experimental full cell. However, it is important to mention that for these high frequencies, the non-ideal inductive behavior already influences the DRT-calculation and is therefore able to influence peak height and peak position. Furthermore, an automatic switch of the measurement range in the frequency response analyzer around 20 kHz distorts the DRT calculation. An unambiguous separation of two physical loss processes is therefore not possible. The DRTs of anode and cathode are rather similar within this frequency range. Therefore it can be stated that both electrodes contribute to the impedance equally.

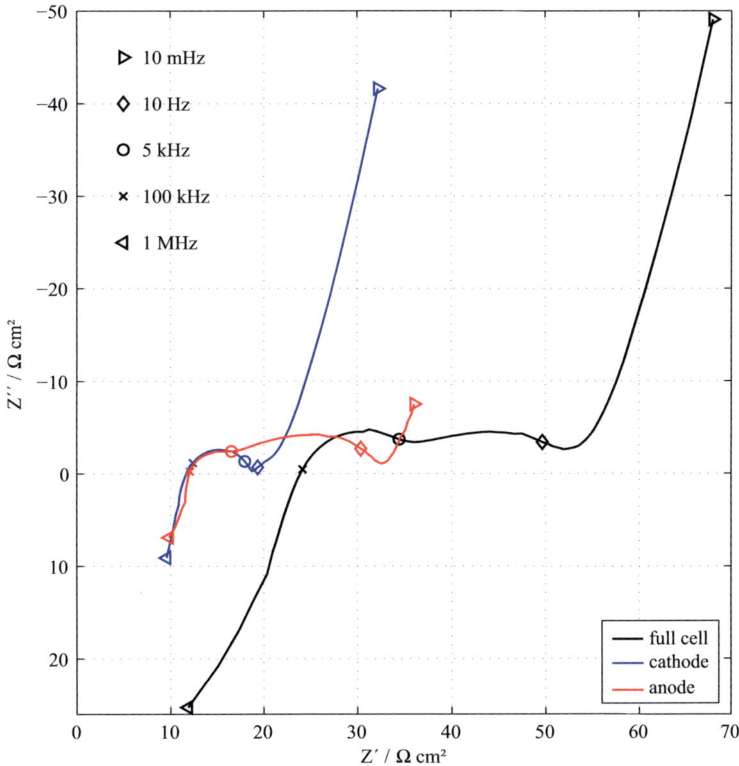

Figure 8.6: Nyquist plot of full cell, cathode and anode impedance spectrum of experimental full cell with reference electrode for SOC=100 % and T=23 °C.

Frequency range II covers frequencies between 10 Hz and 5 kHz. In the DRT, the curve shape of the full cell DRT indicates the existence of up to four characteristic time constants, which is a first hint of either four individual loss processes or less, if one loss process with several coupled time constants takes place here (see Section 2.4). In this case, an inspection of the reference electrode measurements is supportive again, clearly indicating that the anode dominates the impedance contributions in the entire

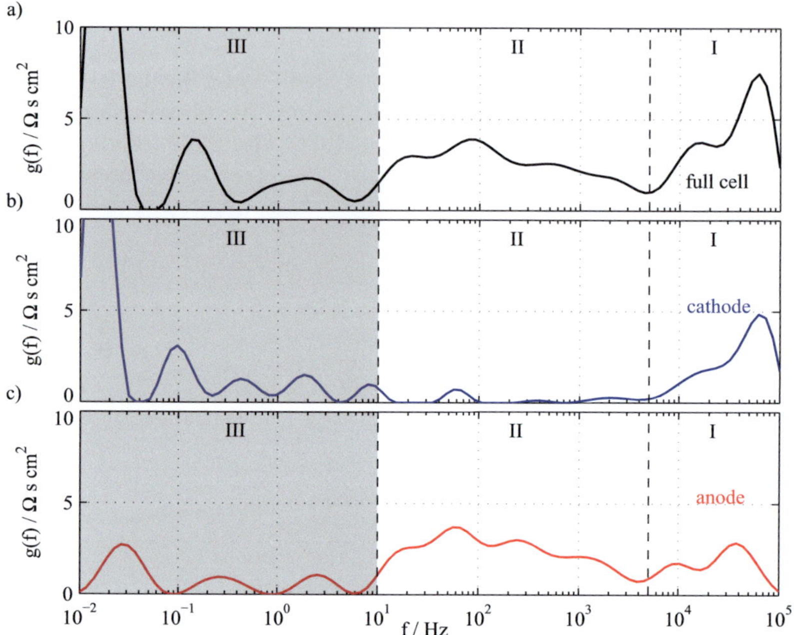

Figure 8.7: DRT for a) full cell, b) cathode and c) anode impedance spectrum of experimental full cell with reference electrode for SOC=100 % and T=23 °C.

frequency range II. The contributions of cathode impedance are negligible.

Frequency range III, assigned to frequencies below 10 Hz, is to be analyzed by time domain measurements. A measurement to lower frequencies than 10 mHz by standard electrochemical impedance spectroscopy is very time consuming and is not suitable for experimental cell measurements as these cells have only limited stability.

8.2.2.2 SOC and Temperature Dependencies of Impedance Spectra

The next thing to be considered is the SOC variation between 100 % and 0 % SOC at T=23 °C, displayed in Figure 8.8 and Figure 8.9. The full cell impedance (Figure 8.8) does not change significantly for SOCs above 10 %. The ohmic resistance and the first semicircle at frequencies above 5 kHz do not show any SOC-dependency at all. The impedance only shows a pronounced SOC-dependency at frequencies below 5 kHz and for very low SOC (10 % and below). A comparison of cathode and anode impedance indicates that the SOC-dependency is caused by the anode. For the cathode impedance, only the capacitive branch changes according to the intercalation capacity (see Section 2.3.2). Another significant dependency on SOC is not visible.

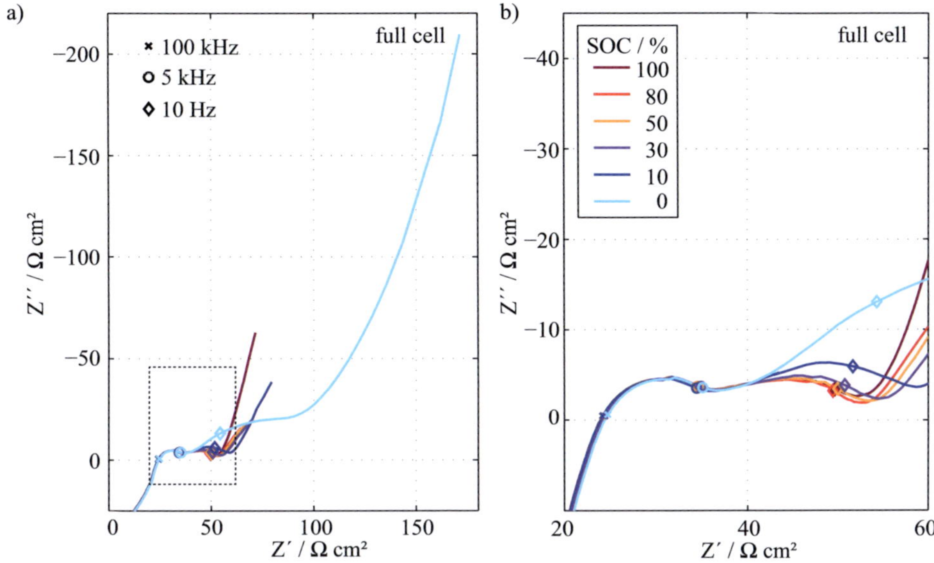

Figure 8.8: Nyquist representation of the impedance spectrum of experimental full cell for varying SOC at T=23 °C. a) Entire spectrum and b) magnification of high frequency region.

Figure 8.10 presents the corresponding DRTs of experimental full cell, cathode and anode for varying SOC. There are merely small variations in the shape of the calculated DRTs for SOCs above 10 % within range II, whereas range I remains totally constant. The DRT changes drastically in range II at SOC=0 %, which corresponds to an open circuit potential of 2.6 V. Here, the predominant peak of the full cell increases significantly in height and its maximum shifts to a frequency more than one order of magnitude lower. The second predominant peak and the decreasing peak sequence for higher frequencies in frequency range II possesses weak SOC-dependency. The measurements via reference electrode in Figure 8.10b reveal that this effect is caused exclusively by the anode, whereas the cathode remains unaffected. Note that the lower border of frequency range II was shifted from f=10 Hz in Figure 8.10 to f=0.5 Hz in order to include the SOC dependent peak.

Figure 8.11 displays the DRTs of experimental full cell, cathode and anode at SOC=100 % for varying temperatures. A weak temperature dependency can be observed in frequency range I. The DRT increases sightly for decreasing temperature. This is mainly caused by the graphite-anode as it can be seen in Figure 8.11c. Once more, frequency range II shows the most distinct parameter-dependency. Both main peaks and the peak sequence at higher frequencies increase steadily with decreasing temperature.

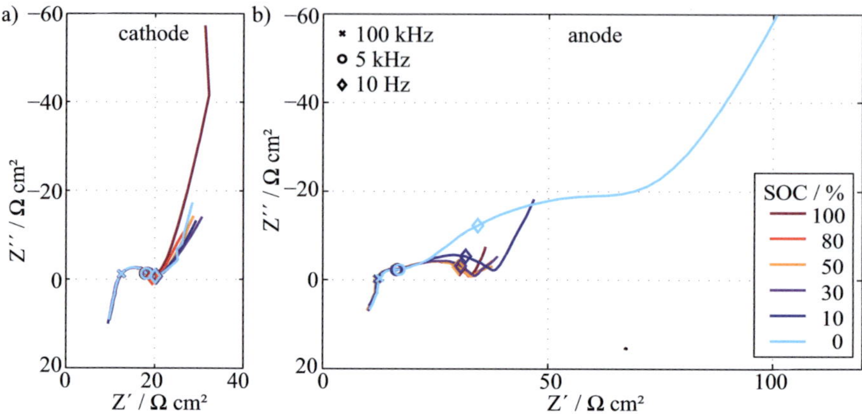

Figure 8.9: Nyquist representation of the impedance spectrum of a) cathode and b) anode measured by reference electrode in an experimental full cell for varying SOC at T=23 °C.

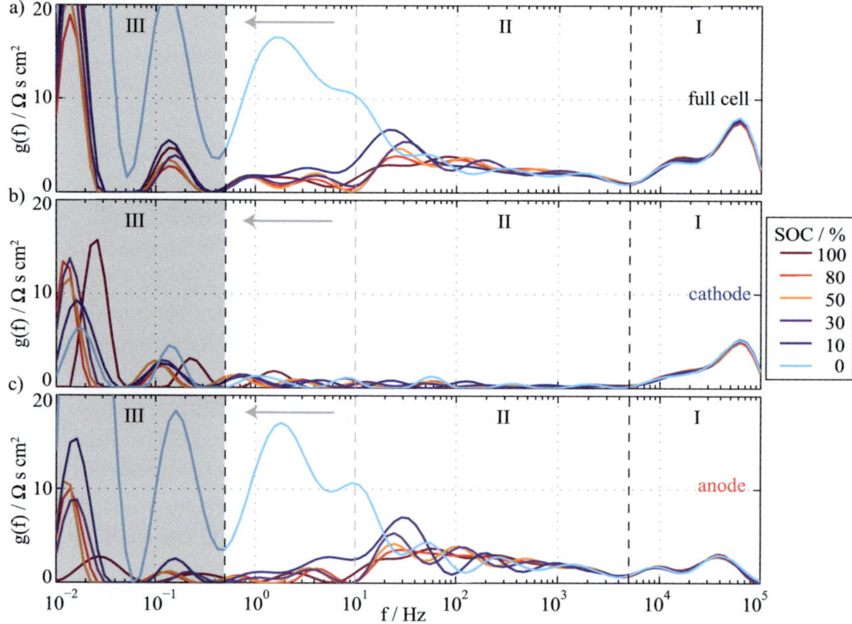

Figure 8.10: DRT of a) full cell, b) cathode and c) anode impedance spectrum for experimental cell with reference electrode for varying SOC and T=23 °C.

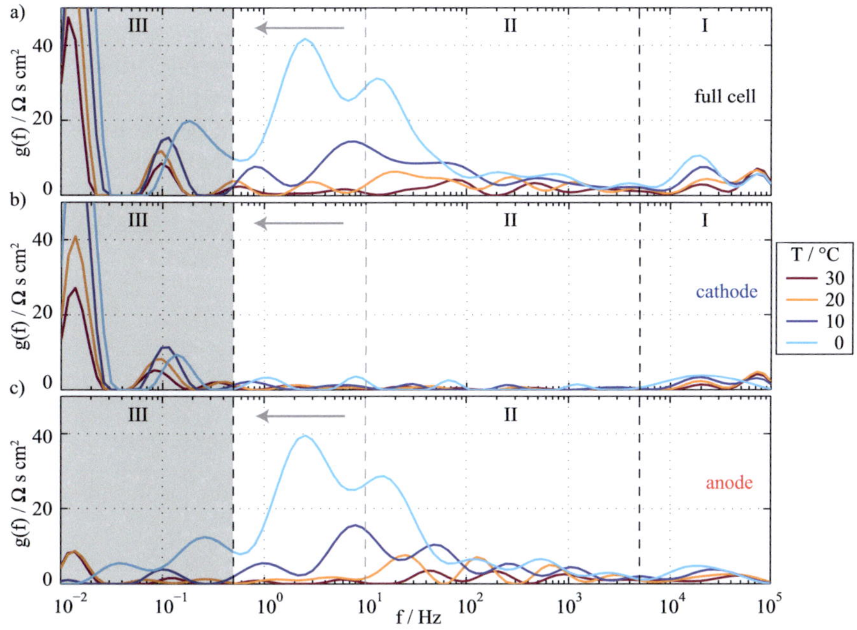

Figure 8.11: DRT of a) full cell, b) cathode and c) anode impedance spectrum for experimental cell with reference electrode for varying temperatures at SOC=100 %.

8.2.2.3 Physical Interpretation of Impedance Spectrum and DRT

The electrodes extracted from the 18650 cell were analyzed in detail in the previous chapters. $LiFePO_4$-cathodes were investigated in Chapter 6, whereas the graphite-anode loss processes were analyzed in Chapter 7. An equivalent circuit model was developed for both electrodes in order to investigate parameter dependencies of each loss process. Furthermore, the physical origin was identified for each loss process in frequency ranges I, II and III. Table 8.3 summarizes the loss processes identified according to their frequency range and their origin and gives their physical interpretation.

Frequency range I contains anode and cathode loss processes of similar polarization. Both were interpreted as contact resistance between electrode layer and current collector. Frequency range II is dominated by the anode losses caused by a combination of charge transfer, transport through the SEI and limited ionic conductivity inside the pores of the electrode. The cathode charge transfer which also takes place in this frequency range does not have a significant impact. The polarization in frequency range III was assigned to the solid state diffusion process in both active materials. A quantitative analysis hereof will be done by TDM. The assignment of each loss process to the peaks in the DRT is illustrated in Figure 8.12. The equivalent circuit models introduced in Chapter 6 and 7 are used in the following section in order to quantify the contribution of each loss process over the entire parameter field.

frequency range	I (100 kHz to 5 kHz)	II (5 kHz to 10 Hz)	III (below 10 Hz)
cathode	$\mathbf{P_{2C}}$ contact resistance cathode / current collector	$\mathbf{P_{1C}}$ charge transfer cathode / electrolyte	$\mathbf{P_{diff,C}}$ solid state diffusion
anode	$\mathbf{P_{CC,A}}$ contact resistance anode / current collector	$\mathbf{P_{CT,A}}$ charge transfer anode / electrolyte $\mathbf{P_{SEI,A}}$ SEI $\mathbf{R_{ion}}$ limited ionic conductivity in the pores	$\mathbf{P_{diff,A}}$ solid state diffusion

Table 8.3: List of identified loss processes and their physical origin. The dominant loss processes for each frequency range are marked with red and blue color. Frequency range III will be analyzed by time domain measurements.

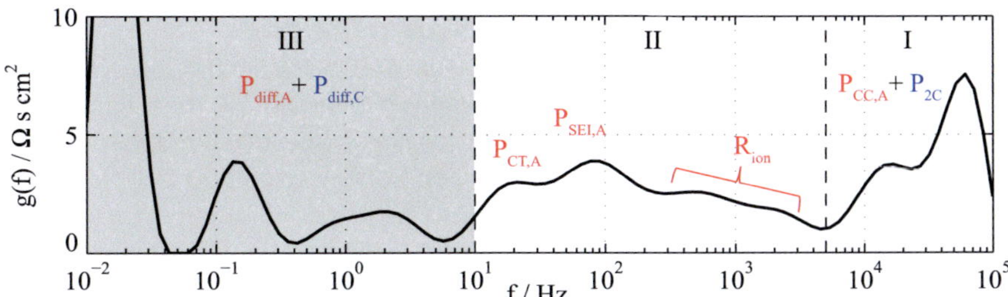

Figure 8.12: DRT of experimental full cell for T=23 °C and SOC=100 % and position of each loss process. The cathode charge transfer P_{1C} is not illustrated as is has a negligible contribution (P_{2C}: Contact resistance cathode/current collector, $P_{CC,A}$: Contact resistance anode/current collector, R_{ion}: Ionic resistance of the electrolyte in the pores, $P_{CT,A}$: Charge transfer anode/electrolyte, $P_{SEI,A}$: SEI anode/electrolyte, $P_{diff,A}$: Solid state diffusion in the anode, $P_{diff,C}$: Solid state diffusion in the cathode).

8.2.2.4 Polarization of Identified Loss Processes

To visualize the performance-limiting factors of the investigated lithium-ion cell, it is necessary to give an overview of all identified loss mechanisms. Therefore, all loss mechanisms identified via impedance spectroscopy and quantified by CNLS-fitting are compared in this section, depending on the operating condition.

Figure 8.13a shows the polarization of all loss processes identified for anode and cathode. Only solid state diffusion is excluded for both electrodes as it is investigated later by TDMs. The anode charge transfer $P_{CT,A}$ is the dominating loss mechanism for low SOCs and low temperatures. It is followed by the lithium-ion transport through the SEI $P_{SEI,A}$ and the contact resistance P_{2C} between cathode and current collector. Anode contact resistance $P_{CC,A}$ and cathode charge transfer P_{1C} play a minor role. The dominating polarizations change for high temperatures and SOCs. The contact resistance between cathode and current collector is then the dominating loss process. The anode charge transfer shows similar polarization whereas all other losses are significantly smaller. Figure 8.14 illustrates the division of polarization into the loss mechanisms for several operation conditions. It illustrates the extremely different contribution of loss mechanisms depending on operation condition. The anode charge transfer $P_{CT,A}$ is the dominant contribution for all operating conditions except for SOC=100 % and T=30 °C. Furthermore, the SEI contribution is larger for SOC=100 % and T=10 °C due to interactions of $P_{CT,A}$ and $P_{SEI,A}$ during the fit procedure. The ASR values of each polarization loss process are listed in Table 8.4. Additionally, the ionic resistance in the anode pores is added in order to show its contribution.

Polarization resistance can be analyzed in more detail by including microstructure parameters in the analysis. It is then possible to understand the origin of the different polarization of interface resistance between cathode/electrolyte and anode/electrolyte better. For this reason, it is necessary to include the active surface area into the analysis. Figure 8.13b shows the polarization of cathode and anode interface loss processes normalized by the active surface area determined by microstructure reconstruction. In contrast to the different absolute contribution of these loss mechanisms to the overall impedance, they show a very similar polarization concerning their active area specific resistance (aASR). The specific active surface area of the anode microstructure is sixteen times smaller than that of the cathode. This is caused by the different average size of LiFePO$_4$ (0.218 μm) and graphite particles (4.236 μm) and it is the main reason for a different order of charge transfer polarization. Such a different order of polarization begs the question, why the graphite-anode is not fabricated with smaller graphite particles. The main reason for the comparably large particle size is the SEI-formation. A larger active surface area in the graphite-anode would also cause a much larger consumption of active lithium during SEI-formation. This would drastically decrease the specific cell capacity and increase the aging due to SEI-formation. Figure 8.13b also reveals the different dependency of both charge transfer resistances on temperature and SOC. The graphite charge transfer $P_{CT,A}$ shows much stronger dependencies than the LFP cathode charge transfer P_{1C}. The temperature dependency in particular is significantly higher. This fact was already shown by the activation energies (0.79 eV for graphite and 0.42 eV for LiFePO$_4$ charge transfer).

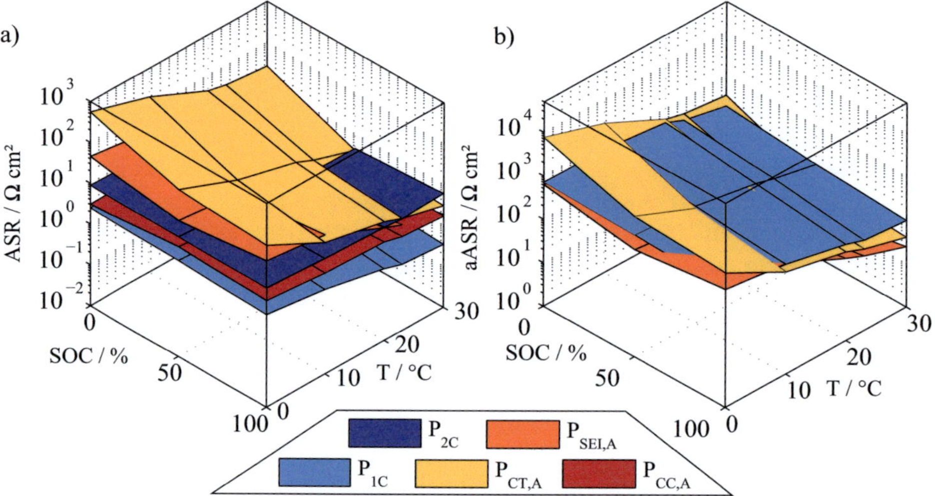

Figure 8.13: a) Area specific resistance for all loss processes identified via impedance spectroscopy and b) active area specific resistance for cathode/electrolyte and anode/electrolyte interface loss processes identified via impedance spectroscopy (P_{1C}: Charge transfer cathode/electrolyte, P_{2C}: Contact resistance cathode/current collector, $P_{CC,A}$: Contact resistance anode/current collector, $P_{CT,A}$: Charge transfer anode/electrolyte, $P_{SEI,A}$: SEI anode/electrolyte, Solid state diffusion in the anode and cathode active material are neglected as they are identified later via TDM).

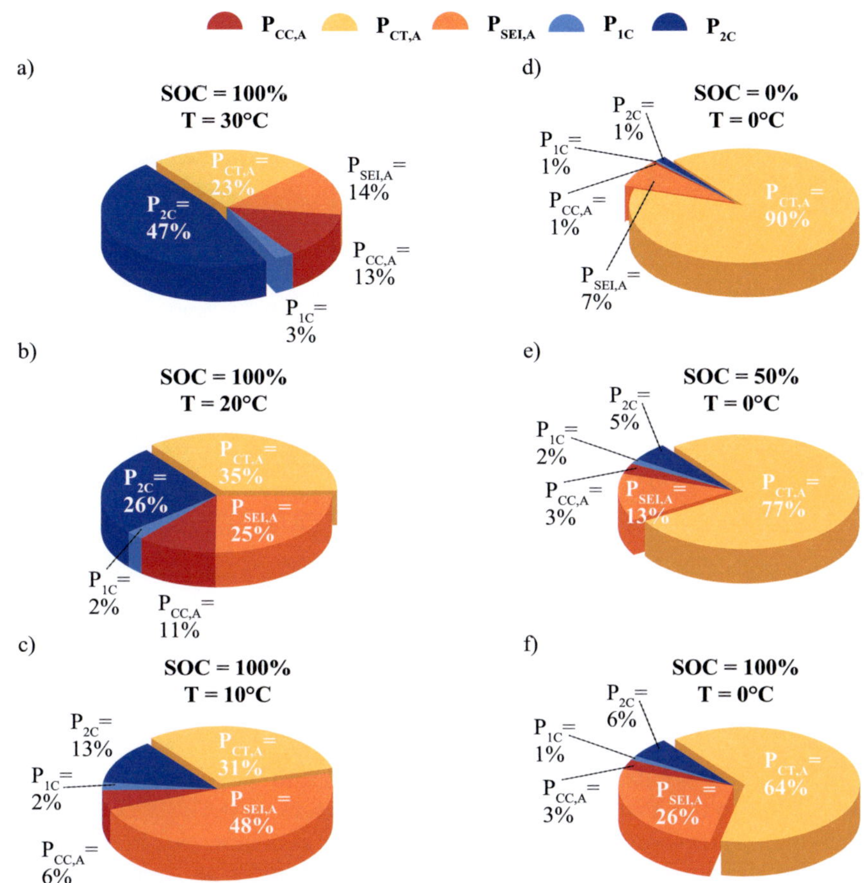

Figure 8.14: Percentage of ASR for all polarization processes in frequency range I and II for *left*: SOC=100 % from T=30 °C to T=10 °C and *right*: T=0 °C from SOC=0 % to 100 % (P_{1C}: Charge transfer cathode/electrolyte, P_{2C}: Contact resistance cathode/current collector, $P_{CC,A}$: Contact resistance anode/current collector, $P_{CT,A}$: Charge transfer anode/electrolyte, $P_{SEI,A}$: SEI anode/electrolyte. Solid state diffusion in the anode and cathode active material are neglected herein, but quantified by using a TDM in Section 8.2.3 et seqq.).

This can be caused by the influence of the SEI on charge transfer resistance. Ref. [43] shows that the temperature dependency of graphite charge transfer is not only dependent on the electrolyte which changes the dissolvation of the lithium. It also depends on the nature of the SEI layer. As the cathode does not form a pronounced reaction-layer (in contrast to the SEI on graphite-anodes), its charge transfer also shows a different temperature activation.

process	ASR / Ωcm^2						physical origin
SOC/ T	100 %/ 30 °C	100 %/ 20 °C	100 %/ 10 °C	100 %/ 0 °C	50 %/ 0 °C	0 %/ 0 °C	
P_{2C}	6.01	5.94	6.87	8.4	8.47	9.03	contact resistance cathode / current collector
P_{1C}	0.34	0.57	1.18	1.87	2.24	2.53	charge transfer cathode / electrolyte
$P_{CC,A}$	1.69	2.59	2.97	4.15	4.9	3.09	contact resistance anode / current collector
$P_{SEI,A}$	1.83	5.8	25.76	38.63	21.42	45.85	SEI anode / electrolyte
$P_{CT,A}$	3.01	8.19	16.57	93.17	123.45	559.61	charge transfer anode / electrolyte
$R_{Ion,A}$	29.22	35.06	42.62	52.55	51.8	59.06	ionic resistance in the anode porosity

Table 8.4: ASR of polarization processes identified for the cathode and the anode in frequency range I and II for various operation conditions in the experimental full cell.

In addition, the SOC-dependency is different for anode and cathode charge transfer. This could be related to the different effect of lithium insertion on the electrode materials. The graphite-anode shows an increasing charge transfer resistance for very low SOC. This means that a very small lithium concentration in the graphite-anode leads to a high charge transfer resistance. A possible explanation for this is the volume expansion of graphite during lithium intercalation. As already explained in Section 2.1.2.2, graphite extends its volume up to 10.3 % when transitioning from the unlithiated to the lithiated state. Assuming a totally empty graphite-anode, this means that the graphene layers have to be expanded in order to insert the first lithium-ions. This insertion and the corresponding expansion cause a high charge transfer resistance for very low lithium concentrations. The volume expansion for $LiFePO_4$-cathodes is only 6.8 % and the $LiFePO_4$ shows a two phase behavior. This leads to other effects which cause the different specific SOC-dependency. The $LiFePO_4$ exhibits a maximum charge transfer resistance for low SOC as well. This corresponds to the fully lithiated state for the cathode active material, indicating that the charge transfer is facilitated when fewer lithium-ions are present in the host lattice.

8.2.3 Identification of Low Frequency Losses via TDM

TDM were performed in order to analyze the low frequency range of the impedance spectrum. For this, experimental full cells without reference electrode were assembled as introduced above because the use of a reference electrode for TDM needs an absolutely stable potential. This is not guaranteed for the reference electrodes previously applied. Therefore, cathode and anode half-cells were analyzed additionally in order to separate the electrode losses reliably. The lithium anode does not play a significant role during time domain measurements, as no intercalation process occurs and therefore the overpotential at the lithium-anode is reduced very fast. A 1C discharge pulse was applied as the excitation signal for $T_p = 10\,\text{sec}$, followed by a voltage relaxation for 30 hours. According to Ref. [74], this leads to a minimum evaluable frequency of $f_{min} = \frac{1}{2\pi\tau_{relax}} = 1.5\,\mu\text{Hz}$. The maximum frequency evaluated due to the limited sample rate is $150\,\text{mHz}$. The measurements were conducted at a full cell SOC of $80\,\%$ at T=25 °C. The corresponding SOC of each electrode needed for half-cell measurements was determined by a $\text{Li}_4\text{Ti}_5\text{O}_{12}$ reference electrode.

8.2.3.1 Evaluation of TDM

The voltage relaxation after a current pulse for an experimental full cell is presented in Figure 8.15. As expected, the voltage increases after the discharge pulse due to the decreasing overpotential. However, the voltage starts to decrease significantly after 7 hours. This phenomenon is denominated as self-discharge and it is detected using the algorithm introduced in Section 2.3.2. The range used for fitting and the fit result of the self-discharge model are illustrated in Figure 8.15. A nearly constant gradient is observed after 17 hours, indicating a linear self-discharge in the full cell. Next, the voltage drop obtained by the model fit is subtracted from the cell voltage. The relevant voltage relaxation due to the decreasing overpotentials and the real OCV remain and they are subsequently analyzed by the pulse-fitting approach.

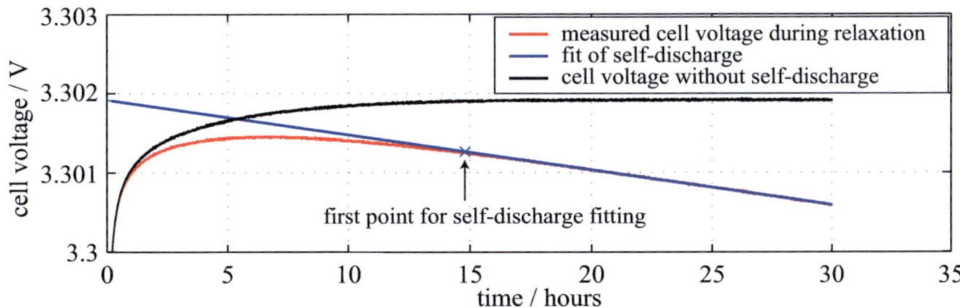

Figure 8.15: Cell voltage of an experimental full cell after a current pulse, fit of the self-discharge model and compensated voltage relaxation for SOC=80 % and T=25 °C.

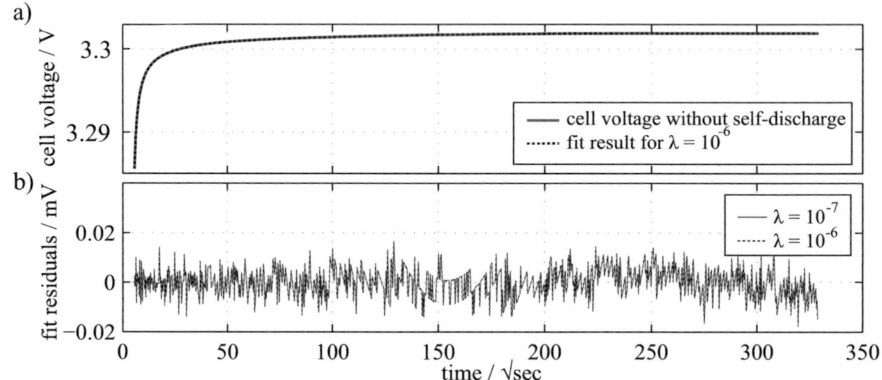

Figure 8.16: a) Compensated cell voltage of an experimental full cell after a current pulse and result of pulse-fitting. b) Residuals of pulse-fitting for two different regularization parameters (SOC=80 % and T=25 °C).

300 RC-elements are chosen to describe the cell voltage. The fit obtained using a regularization parameter of $\lambda = 10^{-6}$ and the consequent residuals are plotted in Figure 8.16. The model is able to describe the cell voltage properly and produces negligible systematic deviations which are smaller than the deviation caused by the measurement noise. The residuals obtained by a smaller regularization parameter are additionally illustrated in Figure 8.16b. Reducing the regularization parameter does not further decrease the systematic deviation, indicating a good parameter choice. Therefore, the results obtained by $\lambda = 10^{-6}$ are evaluated further.

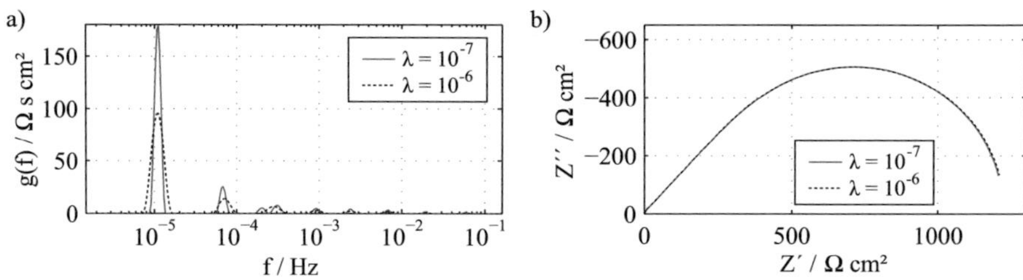

Figure 8.17: a) DRT and b) impedance spectrum of experimental full cell calculated from TDM at SOC=80 % and T=25 °C.

The corresponding DRTs are presented in Figure 8.17a. A decreasing peak sequence with a main peak at 10 µHz is identified in the DRT. Converting the DRT to an impedance spectrum without adding the intercalation capacity $C_{\Delta int}$ results in the Nyquist plot presented in Figure 8.17b. A typical Finite Lentgh Warburg impedance spectrum can

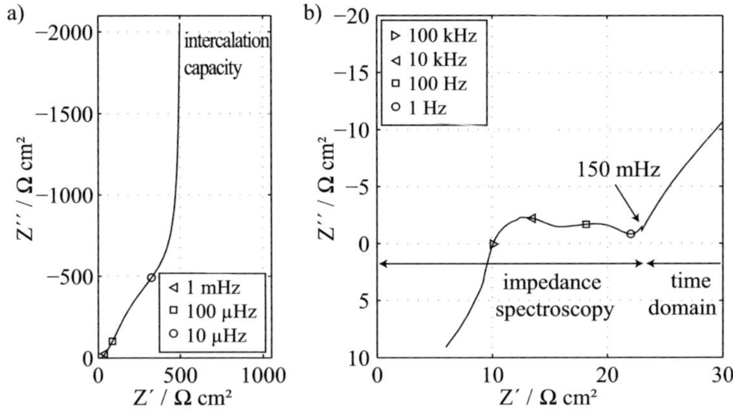

Figure 8.18: a) Merged impedance spectrum of an experimental full cell at SOC=80 % and T=25 °C, calculated from EIS, TDM and intercalation capacity. b) Enlargement of high frequency region and merge point.

be observed, having a constant slope at the high and the shape of an RC-elements at the low frequency range. Furthermore, the impedance spectrum proves that the underlying loss process is fully covered by the measured frequency range because the impedance spectrum converges towards the real axis for the lowest frequency.

The influence of the regularization parameter λ on the DRT is demonstrated in Figure 8.17a as well. A smaller λ results in more narrow peaks, in accordance with the effect of regularization for the standard DRT calculation from impedance spectra. However, the characteristic time constant remains equal and the impedance spectra do not show an influence. As explained above, a regularization parameter of $\lambda = 10^{-6}$ was used for the following analysis.

A merge of the impedance spectra obtained from EIS with those from TDM and the intercalation capacity $C_{\Delta int}$ allows for a visualization of the entire impedance spectrum over almost 12 decades from 1 MHz to 1.5 µHz. Figure 8.18 illustrates the combined impedance spectrum of the experimental full cell. A small intercalation capacity causes the dominant capacitive behavior for low frequencies. Therefore, Figure 8.18b magnifies a part of the impedance spectrum in order to demonstrate the transition between EIS and TDM spectrum. A small deviation between both spectra is observed at the transition frequency around 150 mHz. It is most probably caused by a slight change of the SOC and hence of the impedance spectrum during the voltage relaxation due to the self-discharge of the cell or other aging mechanisms. However, the polarization and the slope of both impedance spectra are consistent at the transition frequency.

8.2.3.2 Separation of Anode and Cathode Losses

Half-cells were measured in order to separate anode and cathode losses reliably. Figure 8.19 compares DRTs and impedance spectra of an experimental full cell and anode and cathode half-cells at SOC=80 % and T=25 °C. It is visible to the naked eye that the cathode causes the major part of the polarization in this frequency range. Full cell and cathode half-cell reveal clearly Warburg-like DRTs and impedance spectra, indicating the solid state diffusion process in the cathode active material to be the physical origin of this low frequency polarization. The DRTs demonstrate a good agreement between full cell and cathode half-cell polarizations as well as time constants. The anode polarization is very small compared to the cathode half-cell, supporting the assumption that only the cathode causes the low frequency losses.

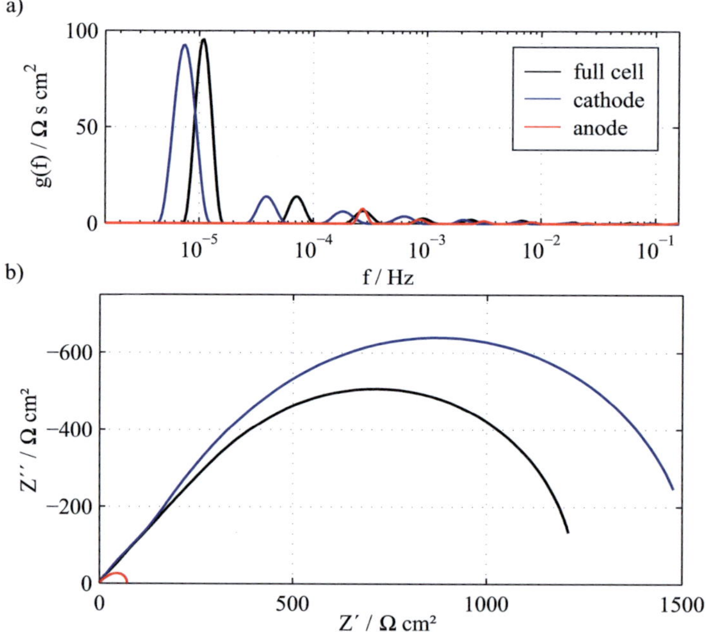

Figure 8.19: a) DRTs and b) impedance spectra of an experimental full cell and half-cells at SOC=80 % and T=25 °C, calculated from TDM without intercalation capacity.

8.2.3.3 Discussion

TDM have been successfully applied in order to determine the low frequency impedance spectrum of experimental lithium-ion cells. The algorithms presented in Ref. [74] for pulse-fitting and self-discharge compensation have been used for the evaluation of these measurements. With full cell and additional half-cell measurements, it was possible to assign the major polarization in frequency range III to the cathode. Furthermore, the Warburg-like impedance spectrum indicates the solid state diffusion in the cathodes' active material as physical origin for this polarization process.

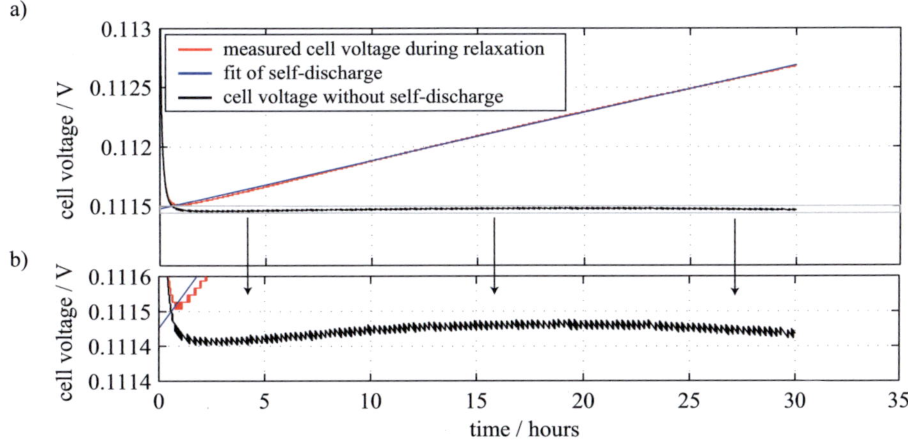

Figure 8.20: a) Voltage relaxation of a graphite half-cell after a current pulse, linear fit of the self-discharge and compensated cell voltage. b) Enlargement of the compensated cell voltage.

A linear self-discharge was assumed for the evaluation of the TDM. However, this does not hold true for all experimental cells. The cathode half-cells did not show any self-discharge, thus representing the best case for TDM analysis. The full cells revealed a linear self-discharge which can be easily compensated for. However, parts of the graphite half-cells showed a non-linear self-discharge as depicted in Figure 8.20. An increasing cell voltage caused by a strong self-discharge can be observed. A compensation with the previously introduced algorithm leads to the cell voltage in Figure 8.20b. It does not show a constantly negative gradient, indicating that the self-discharge was not compensated sufficiently for the first hours and that it changes its gradient. Such a voltage relaxation is not describable by the pulse-fitting approach. This affects the TDM impedance significantly and leads to results that are not meaningful at all. The fact that this kind of non-linear self-discharge only occurs for lithium half-cells indicates first, that it is caused by the SEI formation on the graphite-anode, and second, that during the relaxation of graphite half-cells an SOC between two voltage plateaus is passed. In such an SOC range, the voltage may change non-linearly although a constant self-discharge takes place. This leads to the conclusion that the application of TDM on experimental cells is meaningful only if proper SOCs are chosen. Furthermore, the stability of the experimental cells and the electrolyte used is a critical issue as that influences the intensity of self-discharge.

8.3 Transfer of Experimental Cell Results to the 18650 Cell

The goal of the experimental cell analysis introduced is an investigation of the 18650 cell and its loss mechanisms. Therefore, it is necessary to check the plausibility of the experimental cell results and to check the transfer of the results obtained to the 18650 cell. The comparison is divided into a comparison of EIS and TDM results. It includes a DRT analysis and the application of simplified models for the 18650 cell. The parameters obtained are afterwards compared to the experimental cell results and finally, discussed.

8.3.1 Transfer of EIS Results

8.3.1.1 Direct comparison of EIS and DRT

The first step in order to guarantee a reliable analysis of the 18650 cell by experimental cells is a comparison of polarization and identified loss processes. Figure 8.21a compares EIS measurements of the 18650 cell (green line) with the experimental cell (black line) for 80 % SOC at T=25/23 °C. The corresponding DRTs of both cell types are given in Figure 8.21b. The polarization of the experimental cell measurement is in the same order as the 18650 cell when both curves are normalized on the electrode area. However, the shapes of the impedance spectra differ noticeably. The first difference is that the experimental cell shows a higher ohmic resistance R_0. This results from using another electrolyte and another separator in experimental cells. In particular, the separator in experimental cells is significantly thicker (d=4x220 µm) than the separator in the 18650 cell (d=25 µm). Furthermore, different microstructures for the separators and the use of another electrolyte cause differences for the ohmic resistance.

Secondly, the polarization resistance R_{Pol} is larger for the experimental cell and shows another characteristic shape. The 18650 cell shows one characteristic semicircle, whereas the experimental cell shows two arcs from medium to high frequencies. These characteristics evolve from the different inductance of both types of cells. This can be concluded from the x-axis interception frequency, which is below 5 kHz for the 18650 cell and at 100 kHz for the experimental cell. This demonstrates that the high frequency part of the 18650 cell impedance is governed by the inductance of current collectors and measurement cables, which superimposes the polarization losses taking place at the same frequencies.

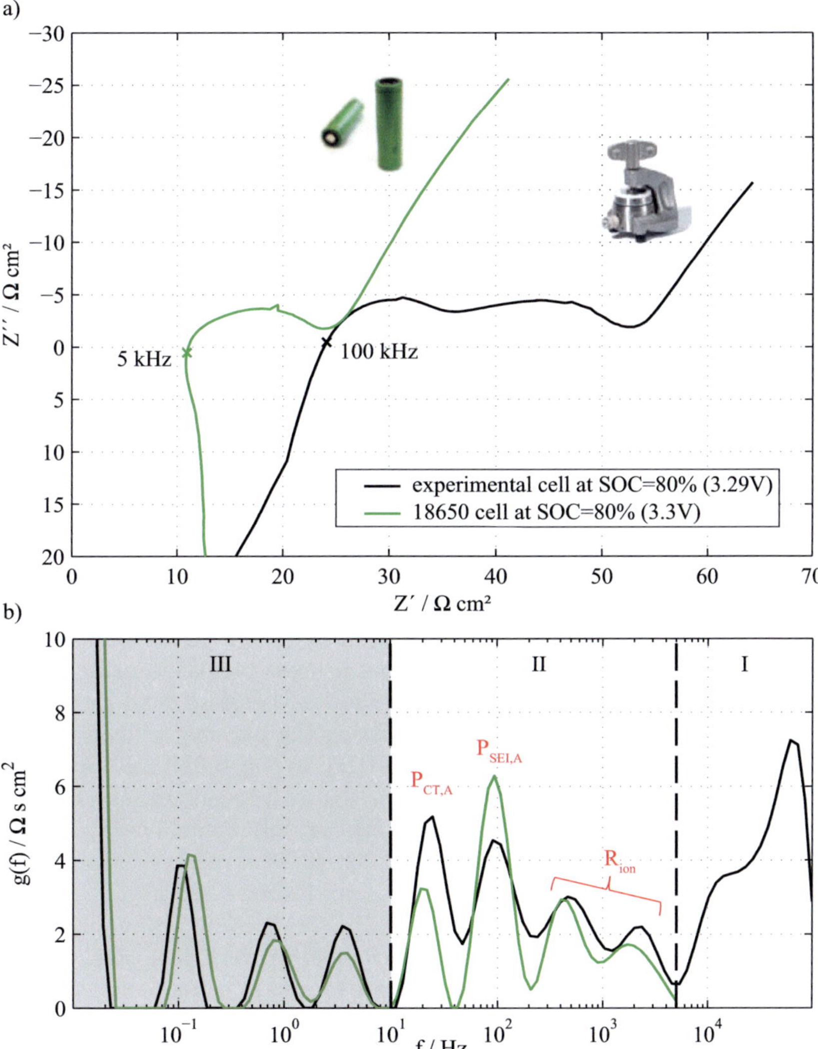

Figure 8.21: a) Nyquist plot and b) DRT of 18650 and experimental full cell impedance spectrum for SOC=80% and T=25/23 °C (R_{ion}: Ionic resistance of the electrolyte in the pores of the anode, $P_{CT,A}$: Charge transfer anode/electrolyte, $P_{SEI,A}$: SEI anode/electrolyte).

More information is gained from the comparison of both calculated DRTs, displayed in Figure 8.21b. Range I is only accessible by experimental cells, as these have a low inductance. As explained in Section 8.2.2, two characteristic peaks are clearly distinguishable and anode and cathode contribute to the impedance significantly. It is deduced that these high frequency contributions occur in the 18650 cell as well. However, they are included in the ohmic resistance R_0 of the 18650 cell. Furthermore, the polarization of these high frequency loss mechanisms is influenced by the preparation of the electrodes. Their contribution to the 18650 cell impedance is therefore expected to be smaller than for the experimental cells.

The DRTs in range II, with frequencies correlating to the second semicircle of the impedance spectrum of the experimental cell, show very good agreement. Both types of cells show four peaks located at the same characteristic frequencies. The lowest frequency peak, interpreted as charge transfer process $P_{CT,A}$, shows lower polarization for the 18650 cell, whereas the second peak, interpreted as SEI process $P_{SEI,A}$, shows larger polarization. This might result from the electrode preparation procedure or the use of a different electrolyte. As discussed in Section 8.2.2, the anode dominates the impedance response over the entire frequency range.

Frequency range III, assigned to frequencies lower than 10 Hz, is interpreted only by time domain measurements in Section 8.3.2.

8.3.1.2 DRT of 18650 Cell at Parameter Variation

In addition to the number and size of polarization of loss processes, it is also helpful to validate the parameter dependency of each loss process identified using experimental cells with a reference electrode. A first comparison is therefore conducted by applying the DRT-analysis on the 18650 cell and identifying the parameter dependency of each separable loss process. Figure 8.22 shows the DRT of the 18650 cell for varying SOC. Only frequency range II can be compared directly, as frequency range I is dominated by the strong inductance and frequency range III can only be analyzed reliably by time domain measurements. Therefore, the direct comparison of experimental and 18650 cells is limited to anode charge transfer $P_{CT,A}$ and anode SEI $P_{SEI,A}$, as these loss mechanisms dominate frequency range II.

The most distinct SOC-dependency can be found for very low SOC (below 10 %). However, the increase of the anode charge transfer peak in frequency range II is not as significant as for the experimental cells (see Figure 8.8). This is caused by the different OCV of 18650 and experimental cell for low SOC. For SOC=0 %, the voltage of the experimental cell (U_{OCV}=2.6 V) is 0.22 V smaller compared to the 18650 cell (U_{OCV}=2.82 V). This difference is due to the strong voltage recovery of the 18650 compared to the experimental cell after reaching the discharge cut-off voltage. The voltage relaxation is caused by additional lithium homogenization processes along the large electrodes. The lithium concentration gradients among the electrode sheets do not occur in experimental cells as their electrode area is significantly smaller.

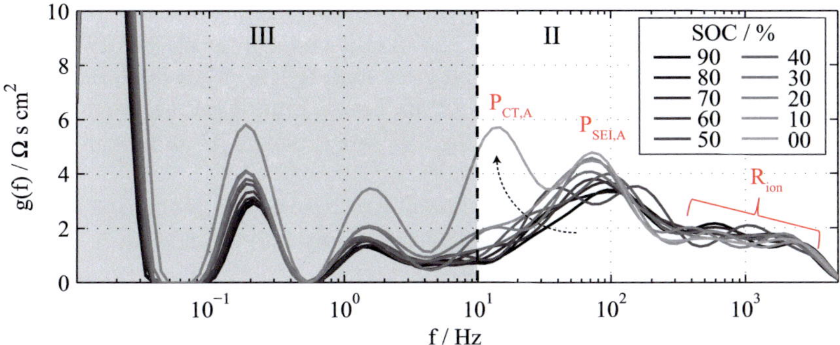

Figure 8.22: DRT of 18650 cell for varying SOC at T=25 °C (R_{ion}: Ionic resistance of the electrolyte in the pores of the anode, $P_{CT,A}$: Charge transfer anode/electrolyte, $P_{SEI,A}$: SEI anode/electrolyte).

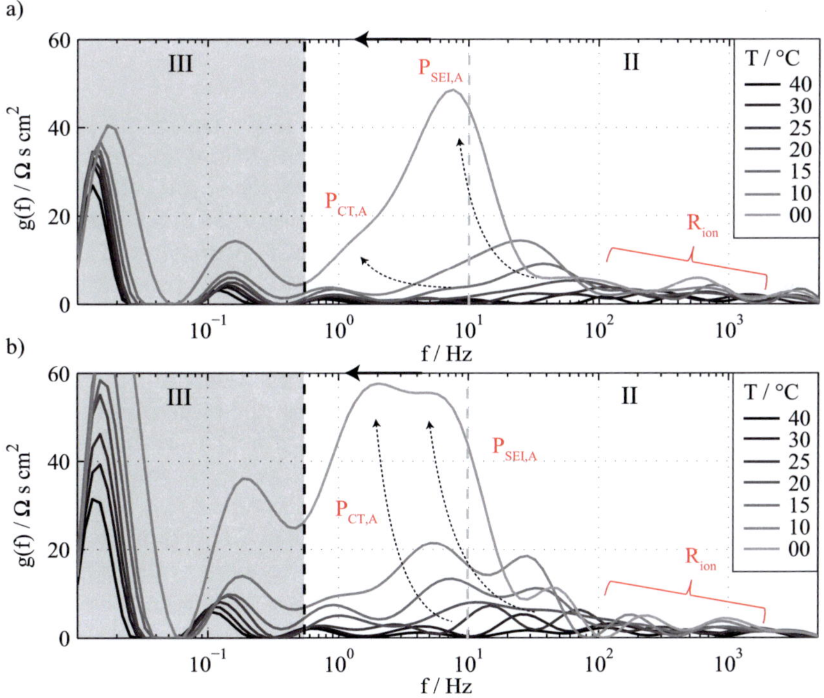

Figure 8.23: DRT of 18650 cell for varying temperature at a) SOC=90 % and b) SOC=0 % (R_{ion}: Ionic resistance of the electrolyte in the pores of the anode, $P_{CT,A}$: Charge transfer anode/electrolyte, $P_{SEI,A}$: SEI anode/electrolyte).

To sum up, the SOC-dependency of the 18650 cell is comparable to the experimental cell. Figure 8.22 indicates that for the 18650 cell the anode charge transfer peak is smaller compared to the SEI-peak, especially for high SOCs. This deviation was already seen in Figure 8.21 and is even more obvious here. Therefore, an easy separation of charge transfer and SEI is not possible for the 18650 cell. The same can be shown for temperature variation.

Figure 8.23a shows the DRTs of the 18650 cell for varying temperatures at SOC=90 %. In contrast to the DRTs of the experimental cells, only one main peak is visible for low temperatures. The previously identified charge transfer process is even smaller compared to the SEI process than for high temperatures and cannot be separated unambiguously. Figure 8.23b shows the DRT of the 18650 cell for varying temperatures at SOC=0 %. This allows for a better visualization of the charge transfer for low temperatures as the separation of charge transfer and SEI is clearer here. A quantitative comparison of parameter dependencies is introduced in the following section.

8.3.1.3 Simplified ECM for 18650 Cell

An exact analysis of loss processes by measuring the 18650 cell is not possible as the high frequency loss processes cannot be assessed and anode and cathode losses overlap for all frequencies. However, it was shown above by experimental cells that frequency range II is dominated by anode contributions for the introduced 18650 cell. Therefore, it is possible to apply a simplified model for the direct evaluation of the 18650 cell in order to asses the parameter dependencies of the anode loss processes in frequency range II.

This simplified impedance model (shown in Figure 8.24) includes an ohmic resistance and an inductor in parallel with an ohmic resistance to describe the high frequency impedance. Here, the serial ohmic resistance represents the electrolyte resistance and the high frequency loss processes of frequency range I which are hidden due to the strong inductance. The inductor in parallel with an ohmic resistance is suitable to describe the inductance of the 18650 cell. This model is very often used to describe a commercial cell's inductance [172, 173] and provides a good fit quality. According to Ref. [51], the skin effect is the physical origin of such RL-type impedance spectrum. Furthermore, the model includes the anode-electrolyte interface resistance, consisting of charge transfer $P_{CT,A}$ and SEI losses $P_{SEI,A}$ integrated into a transmission line model as introduced in Section 7 and the cathode solid state diffusion $P_{diff,C}$. The cathode charge transfer P_{1C} is neglected as it is very small compared to the anode contributions and it therefore cannot be separated in the full cell. Furthermore, the anode solid state diffusion is neglected and described together with the cathode solid state diffusion by a 3D-diffusion model. These simplifications limit the significance of the fit results. Therefore, only a rough comparison between these results and the experimental cell results is possible.

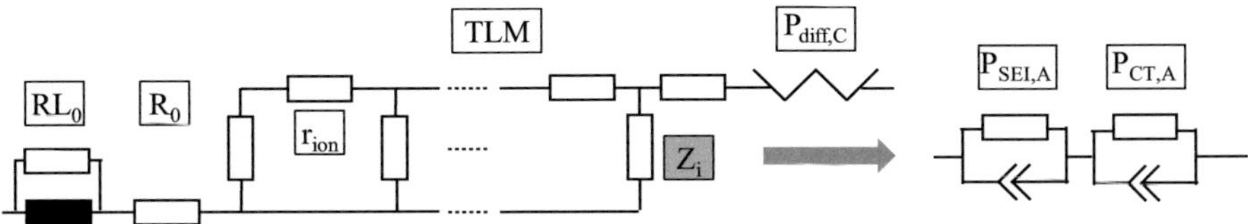

Figure 8.24: Simplified ECM to describe the 18650 cell impedance spectrum, neglecting the high frequency loss processes P_{2C} and $P_{CC,A}$, the cathode charge transfer P_{1C} and the anode solid state diffusion $P_{diff,A}$ (r_{ion}: Ionic resistance of the electrolyte in the pores of the anode, $P_{CT,A}$: Charge transfer anode/electrolyte, $P_{SEI,A}$: SEI anode/electrolyte, $P_{diff,C}$: Solid state diffusion in the cathode, RL_0: inductance model for the 18650 cell, R_0: ohmic resistance).

8.3.1.4 Comparison of Parameter Dependencies

The simplified model will now be used to compare the ASR and the parameter dependencies of anode charge transfer and SEI of the 18650 cell to experimental cells. Figure 8.25a shows the ASR of charge transfer $P_{CT,A}$ depending on SOC. The order of polarization for experimental and 18650 cells is generally the same; however, the polarization of the 18650 cell is slightly smaller. This was already obvious at the DRT analysis and indicates the influence of the different electrolyte used in each cell. Both cells also reveal the similar SOC dependency. The increasing difference at low SOC is caused by the different OCV of both cells. The OCV at 0 % SOC is 2.8 V for the 18650 cell, but it is only 2.6 V for the experimental cell. This causes a stronger effect of increasing polarization at low SOC for the experimental cell. The ASR of $P_{SEI,A}$ is slightly larger for the experimental cell and exhibits the same SOC dependency. The transfer of the results obtained from the experimental cell analysis to the 18650 cell seems to be valid.

The activation energy is another value to compare experimental and 18650 cell measurements. Activation energies were determined for the 18650 cell at SOC=0 % as a clear separation of charge transfer and SEI is only possible for this SOC (see Figure 8.23). The fitting procedure was the same as for experimental cells, assuming an activation energy of 0.14 eV for the ionic conductivity inside the pores. An activation energy of 0.76 eV was determined from these evaluations for $P_{CT,A}$, whereas 0.75 eV was determined for $P_{SEI,A}$ (see Figure 8.26).

Comparing these values to experimental cell results shows a good agreement. The activation energy is slightly lower for $P_{CT,A}$ compared to the experimental cell (0.79 eV), whereas it is higher for $P_{SEI,A}$ (0.72 eV). The reason for this difference may be the use of another electrolyte which affects the temperature dependency of both processes. This fits well with the discussion in Chapter 7, wherein it was stated that the activation energy of charge transfer depends on the solvent used in the electrolyte. Therefore, it can be concluded that the use of an original electrolyte would be desirable to receive comparable results. The fact that this electrolyte is usually not available for the analysis must be considered. Another reason may be an interaction of both loss processes at the CNLS-fit as they have very close time constants which do not allow for a strict separation.

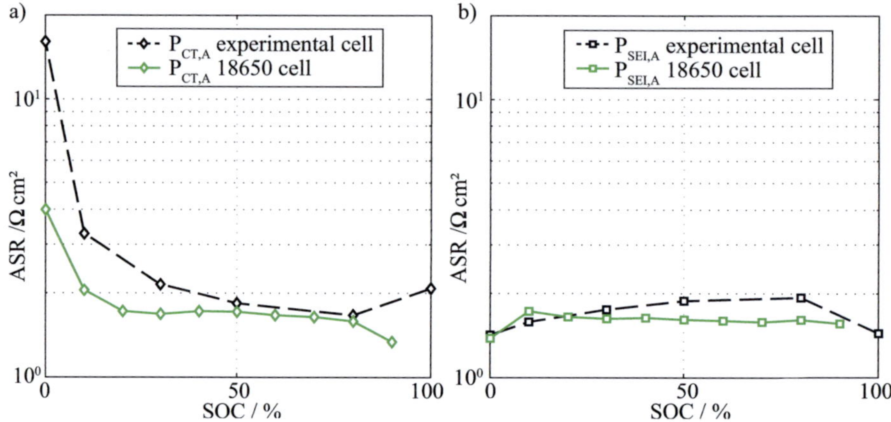

Figure 8.25: ASR of a) anode charge transfer $P_{CT,A}$ and b) SEI losses $P_{SEI,A}$ plotted versus SOC and compared for experimental (T=23 °C) and 18650 cell (T=25 °C). The values for the 18650 cell were obtained from fitting the simplified ECM presented in Figure 8.24.

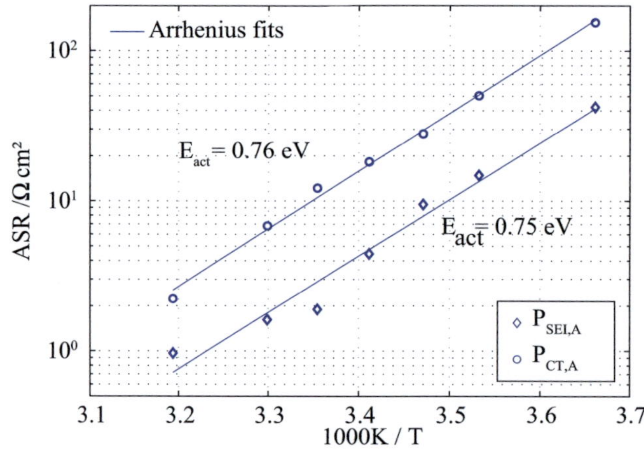

Figure 8.26: ASR of anode charge transfer $P_{CT,A}$ and SEI losses $P_{SEI,A}$ in the 18650 cell for varying temperatures at SOC=0 %. These values were obtained from fitting the simplified ECM presented in Figure 8.24.

8.3.2 Transfer of TDM Results

Finally, the results of the experimental cell analysis in the low frequency region are transferred by comparing the TDM of 18650 and experimental cell. DRT and impedance spectra have been measured for both cells for SOC=80 % at T=25 °C. Figure 8.27 presents the comparison of both cells. The DRT can be split up in two characteristic frequency ranges. Range III covers frequencies above 5 µHz. DRT and impedance spectrum of the experimental cell point to a Warburg type impedance in this frequency range, indicating the solid state diffusion in the active material as the physical origin. As shown by half-cell measurements, the cathode is most relevant for this polarization contribution. Therefore, the solid state diffusion taking place in the $LiFePO_4$-cathode of the 18650 cell is the dominant polarization loss over the entire frequency range III. However, the DRT of the 18650 cell reveals at least one additional peak for frequencies bellow 5 µHz which are denominated as frequency range IV. This peak was not identified for the experimental cells. Hence, this polarization process remains nonassignable to anode or cathode, as this DRT peak (a) does not exist in the experimental cell or (b) cannot be measured due to stability problems at these very low frequencies. Assuming cause (a), the physical origin may be related with a homogenization process inside the entire volume of the electrode layers. Probably, there are uncovered lithium concentration gradients among the electrodes of the 18650 cell. Naturally, such concentration gradients do not exist in the experimental cells which have only a very limited electrode area.

8.4 Discussion and Conclusions

This chapter focused on the analysis of the commercially available 18650 cell addressing three main goals:

1. Increase of the evaluable frequency range to higher and lower frequencies.
2. Superior separation of overlapping loss processes.
3. Physical interpretation of the 18650 cell impedance spectrum.

First, the 18650 cell was opened and its electrodes extracted. This procedure enabled the measurement of the latter in experimental cell housings. Reproducibility is a critical issue for the electrodes extracted due to the removal of one side of the double-side coated electrodes. The influence of different preparation methods on capacity and impedance spectra was analyzed in order to chose the optimal preparation method. H_2O turned out to be the best solvent for cathode preparation as it provided a good compromise between high capacities and acceptable reproducibility of the impedance spectra. An influence of this procedure on the electrodes cannot be avoided completely. However, two advantages result from the analysis of the electrodes in experimental cell housings. First, a separation of cathode and anode losses via reference electrode is possible. Second, the frequency range can be expanded to higher frequencies by almost two decades (to 100 kHz) due to the lower impact of cable and cell inductance.

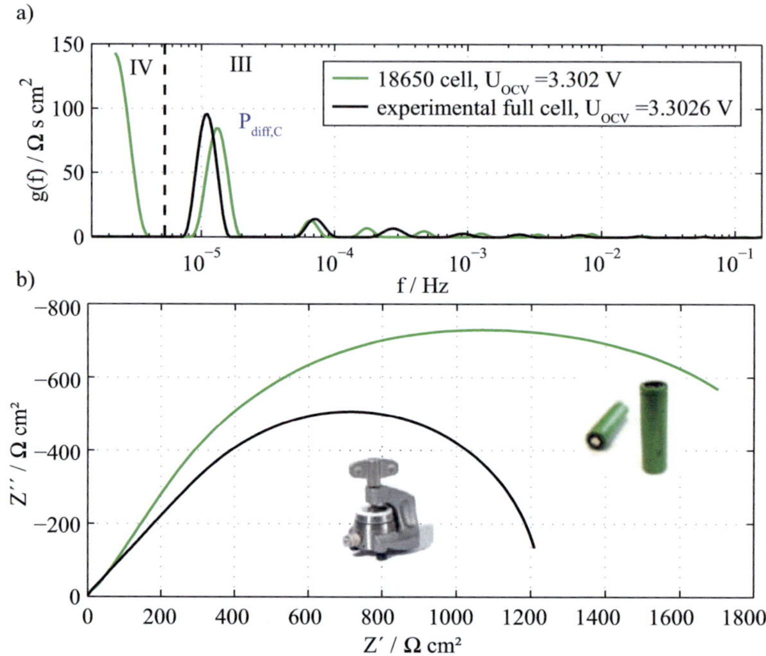

Figure 8.27: a) DRT and b) impedance spectrum of an experimental and a 18650 full cell, calculated from time domain measurements at T=25 °C and SOC=80 % ($P_{diff,C}$: Solid state diffusion in the cathode active material).

The measured impedance spectra were subsequently evaluated by DRT which increased the separability of loss processes significantly. The physically motivated equivalent circuit models, obtained for both electrodes by DRT analysis in Chapter 6 and 7, were applied in order to determine the polarization of all occurring loss processes dependent on operation parameters. This gave an overview of all contributions in a wide parameter field. Next, these results were transferred to the 18650 cell. This was done by comparing the area specific DRTs and, for frequency range II, by comparing the ASR of both relevant loss processes.

The low frequency range was expanded by conducting additional TDM on experimental and 18650 cells. It was demonstrated that the application of TDMs on experimental cells and with this the separation of anode and cathode losses is possible. The results could be transferred to the 18650 cell successfully because the TDM impedance only differed in the scaling factor of electrode area. Furthermore, an additional loss process was identified in the 18650 cell.

The detailed results for each frequency range are:

1. Range I from $100\,\mathrm{kHz}$ to $5\,\mathrm{kHz}$ (measured by EIS): Frequency range I could be analyzed by experimental cells, providing information about additional high frequency loss processes such as contact resistances (P_{2C}, $P_{CC,A}$). The ASR of these loss processes is influenced by the extraction and preparation of the electrodes, causing a certain variation of the ASR in experimental cells. This means that for the 18650 cell, the ohmic resistance consists of several contributions, including electrolyte resistance, contact resistances and electronic conductivity of electrodes and current collectors. The quantitative contribution of each part cannot be determined exactly. Having the original electrolyte from the 18650 cell would allow for an exact division of components. The original electrolyte was not available for the 18650 cell investigated, as no surplus electrolyte could extracted.

2. Range II from $5\,\mathrm{kHz}$ to $10\,\mathrm{Hz}$ (measured by EIS): For frequency range II, two anode loss processes ($P_{SEI,A}$ and $P_{CT,A}$) in combination with the limited ionic conductivity in the electrolyte were identified as main contributions. Since this frequency range is also accessible by 18650 cell measurements and the cathode contribution by the charge transfer process P_{1C} is negligibly small, it was possible to compare the experimental cell results directly to the 18650 cell. The ASR of experimental and 18650 cells fits very well, although the relation between $P_{SEI,A}$ and $P_{CT,A}$ is slightly different. The characteristic SOC-dependencies of $P_{SEI,A}$ and $P_{CT,A}$ are the same for both cells, whereas the temperature dependencies are slightly different. The difference may result from CNLS-fitting or due to the different electrolyte as the electrolyte influences the temperature dependency of charge transfer and SEI. Again it would be desirable to use the original electrolyte from the 18650 cell in order to get undistorted results.

3. Range III from $10\,\mathrm{Hz}$ to $5\,\mathrm{\mu Hz}$ (measured by TDM): TDM were successfully applied on experimental cells in order to determine their low frequency impedance spectrum. The self-discharge turned out to be non-linear for some graphite half-cells, causing problems with the algorithm applied for compensation. The evaluation of experimental fulls cells and $LiFePO_4$ half-cells was straightforward, as only a linear or even no self-discharge occurred. The measurements of full cells and $LiFePO_4$ half-cells demonstrated that the cathode causes the main contribution in frequency range III. The anode does not contribute significantly. The obtained impedance spectra clearly describe a Warburg-like impedance spectrum, indicating that a diffusion process is the physical origin. These results were transferred to the 18650 cell by comparing the area specific low frequency impedance spectra. Comparable polarizations and time constants were obtained for the experimental full cell and the 18650 cell in frequency range III, enabling a direct transfer of the results previously introduced. The solid state diffusion taking place in the $LiFePO_4$-cathode ($P_{diff,C}$) of the 18650 cell is the dominant polarization loss over the entire frequency range. This is an unexpected result; however, the anode polarization ($P_{diff,A}$) seems to be negligibly small.

4. Range IV below $5\,\mathrm{\mu Hz}$ (measured by TDM): No loss process was identified for experimental cells below $5\,\mathrm{\mu Hz}$. However, the DRTs of the 18650 cell reveal

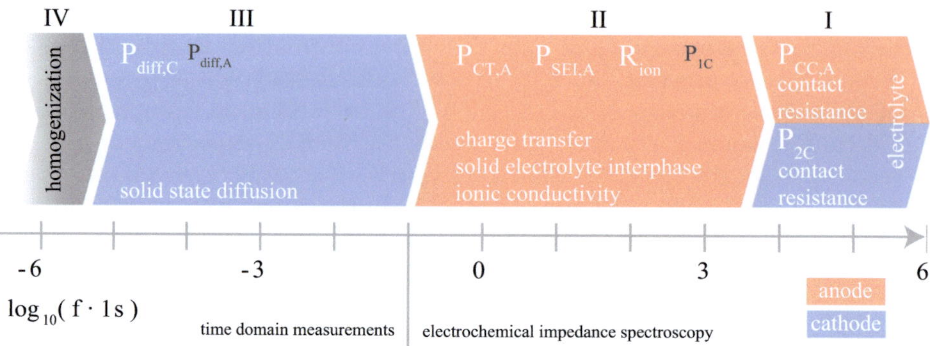

Figure 8.28: Overview of dominant contributions identified for the 18650 cell according to their characteristic frequency range. The cathode charge transfer P_{1C} and the anode solid state diffusion $P_{diff,A}$ are shaded in gray as they were identified in the experimental cells but contribute negligibly to the 18650 cell impedance spectrum (P_{2C}: Contact resistance cathode/current collector, $P_{CC,A}$: Contact resistance anode/current collector, P_{1C}: Charge transfer cathode/electrolyte, R_{ion}: Ionic resistance of the electrolyte in the pores, $P_{CT,A}$: Charge transfer anode/electrolyte, $P_{SEI,A}$: SEI anode/electrolyte, $P_{diff,A}$: Solid state diffusion in the anode, $P_{diff,C}$: Solid state diffusion in the cathode).

> an additional loss process in this frequency range. Probably, the impedance response of the 18650 cell matching with such a time constant is related to a homogenization process. It is possible that lithium concentration gradients among the entire volume of the electrodes of the 18650 cell are taking place here.

An overview of all dominant contributions according to their characteristic frequency range is given in Figure 8.28.

Comparing the results presented in this chapter to data in the literature shows that such a detailed analysis of the impedance spectrum of commercially available cells does not exist. However, there are several studies dealing with the opening of lithium-ion cells. Abraham et al. [129] present the postmortem analysis of an NCA/graphite cell in experimental cell setup. The evaluation of the impedance spectra obtained does not go beyond a comparison of the anode and cathode impedance spectrum and their entire contribution to the overall resistance.

In Ref. [131], the opening of a commercial cell with similar electrodes compared to this study ($LiFePO_4$/graphite) was presented. The impedance spectra measured for both electrodes are comparable in polarization and also in their characteristic shape as far as a comparison of impedance spectra by the naked eye is meaningful. It is obvious that the DRT analysis applied in this thesis allows for a superior evaluation of the impedance spectra. The advantages are based on a better separation of closed time constants and, even more importantly, on a separate analysis of each frequency range. This enables the individual, frequency dependent determination of dominant contributions.

Furthermore, it was demonstrated in this chapter that the low frequency range can be analyzed in like manner. For this purpose, the standard impedance spectroscopy can be complemented by time domain measurements which are subsequently also evaluated by the DRT approach. Applying these methods to the aged cells presented in Ref. [131] would produce a much deeper understanding of aging mechanisms. Therefore, the application of the approach presented in this chapter on aged commercial cells is highly recommended.

9 Summary

One of the main obstacles for a sustainable success of electric mobility is the limited energy- and power density of todays' energy storage technologies which lead to short cruising ranges and reduced performance for electric vehicles. The use of lithium-ion batteries promises the best prospect of success affording a reasonable but still inadequate performance. However, one main risk for the extensive application of lithium-ion batteries in electric vehicles is their unknown aging behavior, leading to high costs in the case of unexpectedly high performance drop.

Understanding the electrochemical loss mechanisms in lithium-ion cells and their electrodes is a crucial factor for the improvement of current cell generations' performance and lifetime. One of the first aims of this thesis was to identify these loss mechanisms and their physical origin for lab-scale cathodes and to apply this knowledge for a subsequent improvement of the latter. Secondly, anodes extracted from commercially available 18650 cells were analyzed in order to obtain a comprehensive understanding of the interaction between electrochemical loss mechanisms, electrode microstructure and the corresponding impedance spectrum. Lastly, the results from experimental cell measurements in the previous chapters were transferred to an 18650 cell. This transfer led to a physical interpretation of the corresponding impedance spectra which can be applied for future cell improvement or evaluation of its aging mechanisms.

Modelling of Lab-Scale Cathodes

In Chapter 5, experimental half-cells with $LiFePO_4$-cathodes and lithium anodes were investigated by Electrochemical Impedance Spectroscopy (EIS) under systematic variation of operating parameters such as temperature and State Of Charge (SOC). The resulting impedance spectra were then analyzed by the Distribution of Relaxation Times (DRT), a method developed at the IWE for the analysis of solid oxide fuel cells' impedance spectra. Due to the different nature of lithium-ion cells' impedance spectra it was necessary to pre-process the latter in order to obtain a meaningful analysis. This procedure and additional measurements of symmetrical cells allowed for the identification of all relevant loss mechanisms and their physical interpretation. Three major loss processes have been identified for lab-scale $LiFePO_4$-cathodes: (i) The solid state diffusion $P_{diff,C}$ inside the active material, (ii) the charge transfer P_{1C} between active material and electrolyte and (iii) the contact resistance P_{2C} between cathode and current collector.

A physically based equivalent circuit model which is able to represent the cathode and anode loss mechanisms adequately and to quantify their polarization for temperature and SOC variation has been developed. Among the cathode losses, the charge transfer process P_{1C} reveals the strongest dependence on state of charge and temperature, revealing a high activation energy of 0.45 eV. The contact resistance P_{2C} is almost independent from both parameters with an activation energy of 0.06 eV, whereas the solid state diffusion exhibits a dependency on both parameters.

The approach including the DRT presented in this chapter has proven to be a powerful tool for the evaluation and impedance model development of lithium-ion cell electrodes. Therefore it was necessary to use half-cells and symmetrical cells for clear separation of loss mechanisms. This approach can be transferred to the analysis of any kind of lithium-ion cell electrode.

Optimization of Lab-Scale Cathodes

In Chapter 6, the dependence of the polarization of cathode loss mechanisms on electrode microstructure was analyzed by a systematic variation of fabrication parameters such as (i) layer thickness, (ii) material fractions and (iii) an additional calendering step. A new cell configuration using a reference electrode was developed for this analysis in order to obtain more precise results without any influence of the lithium counter electrode. The effects observed for the microstructure variation have proven the physical interpretation of the loss mechanisms and the equivalent circuit model. A linear correlation was found between the active surface area and the charge transfer resistance, whereas a high carbon black content in combination with a calendering step allowed for a decrease of the contact resistance.

Furthermore, these cathodes were analyzed by discharge experiments in order to check the influence of microstructure variation on the electrode performance under load. These results were compared to cathodes extracted from commercially available cells in order to have a relevant benchmark. It was possible to improve the lab-scale cathodes significantly, yet the performance of the benchmark cathode was not reached. However, it was possible to demonstrate the applicability of equivalent circuit modelling for the evaluation of electrode performance. Only the contact resistance between cathode and current collector could be identified as an important contribution. Besides that, the limited electron transport through the cathode layer was a deciding factor at high discharge rates which could not be identified by impedance spectroscopy but could be identified by discharge experiments. Therefore, the combination of both measurements has been found to be a profitable approach for the optimization of an electrodes' performance.

For future optimizations of electrodes with higher area-specific capacities and lower ASR it is necessary to improve the measurement setup as the lithium anode and the electrolyte start to dominate the discharge behavior in such cases.

Modelling of Complex Anode Structures

In Chapter 7, the impedance analysis previously described was applied on graphite-anodes. They were measured in experimental cells with a reference electrode which enabled a direct separation of working and counter electrode. The anodes were extracted from commercially available high power 18650 cells in order to understand the interaction between electrochemical loss mechanisms, electrode microstructure and the corresponding impedance spectrum for state-of-the art anodes. The latter have a significantly higher layer thickness and lower porosity compared to the lab-scale electrodes introduced above.

Applying the DRT analysis, microstructure reconstructions and the theory of Transmission Line Models (TLM) indicated a strong interaction between the occurring loss processes and the properties of the the the porous layer. Four loss processes have been identified in the impedance spectrum: (i) The solid state diffusion inside the active material, (ii) the charge transfer and (iii) the SEI between anode active material and electrolyte, and (iv) the contact resistance between the anode and the current collector. Additionally, the ionic conductivity in the pores was found to contribute strongly to the polarization resistance of the impedance spectrum. From this knowledge, an equivalent circuit model was developed which employs a TLM structure in order to describe the interaction of interface loss processes and limited ionic conductivity properly. Including this interaction in the model was not only important for the determination of the correct polarization, but also for a meaningful evaluation of parameter dependencies of the underlying loss processes. It was demonstrated that such a physically meaningful analysis reveals a higher resistance of the electrolyte inside the pores than expected from microstructure reconstruction and intrinsic conductivity. This result indicates that either the pore volume is decreased compared to the microstructure reconstruction or the ionic conductivity is reduced due to interactions with the electrode material.

The model development and subsequent CNLS-fitting enabled the quantification of all contributions over a wide range of operating conditions (T=0 °C to 30 °C, SOC=0 % to 100 %). The ionic conductivity inside the pores causes a main part of the polarization for higher temperatures (20 °C to 30 °C) and higher SOC (above 10 %). The interface loss processes such as charge transfer $P_{CT,A}$ and solid electrolyte interphase $P_{SEI,A}$ dominate for lower temperatures and SOC.

Finally, the model was transferred to a high energy anode having an even thicker layer coating and a drastically lower porosity compared to the high power anodes previously investigated. The model was able to describe the anode impedance spectrum and to separate the single loss process contributions as well. The contribution of the ionic path through the pores of the electrode was much larger compared to the high power anode as it was expected due to its different microstructure. These results demonstrated on the one hand side the general validity of the equivalent circuit model for different graphite anodes. Furthermore, they emphasized the importance of considering the microstructural impact for a meaningful analysis of most technically relevant anodes by impedance spectroscopy.

Multi-Step Approach for the Analysis of 18650 Lithium-Ion Cells

A detailed analysis of commercially available 18650 cells by using the knowledge gained previously was presented in Chapter 8. The goal was the identification of dominating internal losses in these cells and their components. This knowledge is indispensable for a systematic advancement of lithium-ion cell technology concerning energy and power densities and for a detailed analysis of their aging behavior. The standard impedance analysis of 18650 cells offers only a poor insight into the cells which is caused by several factors:

1. Dominant inductance for high frequencies due to cell geometry and a small polarization.
2. Overlap of anode and cathode losses for all frequencies due to limited frequency resolution and a missing reference electrode.
3. Low frequency limit due to drastically increasing measurement time.

The first step in this approach was to open the cell in order to investigate its components in experimental cell housings. This procedure makes high demands concerning preparation and reproducibility for the extracted electrode sheets. However, the high frequency region could be enhanced by two decades for the impedance analysis due to a lower inductance at the experimental cell measurements. Furthermore, a new type of reference electrode was introduced in order to separate anode and cathode loss processes reliably. Next, the low frequency limit was extended by complementing the standard impedance spectroscopy by time domain measurements which were developed at IWE for the analysis of commercial lithium-ion cells. The application of time domain measurements on different experimental cell configurations was demonstrated and enabled the separation of anode and cathode losses. An expansion of the low frequency limit by four decades could be achieved with this procedure. Altogether, the standard frequency range of commercially available lithium-ion cell analysis from around $5\,\mathrm{kHz}$ to $10\,\mathrm{mHz}$ could be enlarged to frequencies between $100\,\mathrm{kHz}$ and $1\,\mathrm{\mu Hz}$. Finally, the frequency selectivity of loss processes could be increased by calculating the DRT from the gained impedance spectra. The entire procedure enabled the identification and quantification of 7 polarization loss processes for the 18650 cell over a large field of operation conditions:

1. Range I from $100\,\mathrm{kHz}$ to $5\,\mathrm{kHz}$ (measured by EIS): Inductive behavior, most probably originating from the internal current collectors and the measurement cables, dominates the impedance response of the 18650 cell in this frequency range. Experimental cell measurements disclosed that the electrolyte contribution to the ohmic resistance of the 18650 cell is overestimated, as not only the electrolyte as usually assumed [51], but also polarization losses originating from contact resistances anode/current collector ($P_{CC,A}$) and cathode/current collector ($P_{2,C}$) contribute to the ohmic resistance, and do not change with SOC.
2. Range II from $5\,\mathrm{kHz}$ to $10\,\mathrm{Hz}$ (measured by EIS): The anode dominates the impedance response of the 18650 cell over the entire frequency range. The detailed impedance analysis in Chapter 7 identified a mixed contribution of the SEI ($P_{SEI,A}$) and the charge transfer ($P_{CT,A}$) reaction in combination with a limited ionic conductivity in the pores to be major contributions. The low frequency peak assigned to charge transfer increases strongly with decreasing SOC. The charge transfer process between cathode and electrolyte ($P_{1,C}$) which also occurs in this frequency range is negligible.
3. Range III from $10\,\mathrm{Hz}$ to $5\,\mathrm{\mu Hz}$ (measured by TDM): The solid state diffusion taking place in the $LiFePO_4$-cathode ($P_{diff,C}$) of the 18650 cell is the dominant

polarization loss over the entire frequency range. This is an unexpected result; however, the anode polarization ($P_{diff,A}$) seems to be negligibly small. This frequency range was only analyzed for 80 % SOC.

4. Range IV below 5 µHz (measured by TDM): Probably, the impedance response of the 18650 cell matching with such a time constant is related with a homogenization process. It is possible that lithium concentration gradients among the entire volume of the electrodes of the 18650 cell are taking place here. This frequency range was only analyzed for 80 % SOC.

This approach provides a new and comprehensive understanding of the impedance response of commercially available cells. Not only the separation of anode and cathode contributions, but also the interpretation of the ohmic resistance and the consideration of the ionic conductivity in the pores as major contributions are relevant for a correct evaluation.

In a next step, the approach introduced in this thesis will be applied for the analysis of aging mechanisms in automotive cells. A meaningful and correct physical interpretation is necessary in order to draw correct conclusions from aging experiments. It is a main factor in understanding degradation completely and – in the future – to fulfill the requirements of a reliable application of lithium-ion technology in electric vehicles.

10 Appendix

A Commercially available Lithium-Ion Cell used in this Thesis

cell manufacturer	Sony
cell type	SE US18650FT
cell case	18650
electrodes	$LiFePO_4 \,/\, C_6$
nominal capacity	1.1 Ah
charge cutoff voltage	3.6 V
discharge cutoff voltage	2.0 V
charge protocol	CC(1C) CV(1h C/20)

B Opening of 18650 Cells

This section illustrates the opening of 18650 cells and the subsequent preparation of double-side coated electrodes in pictures.

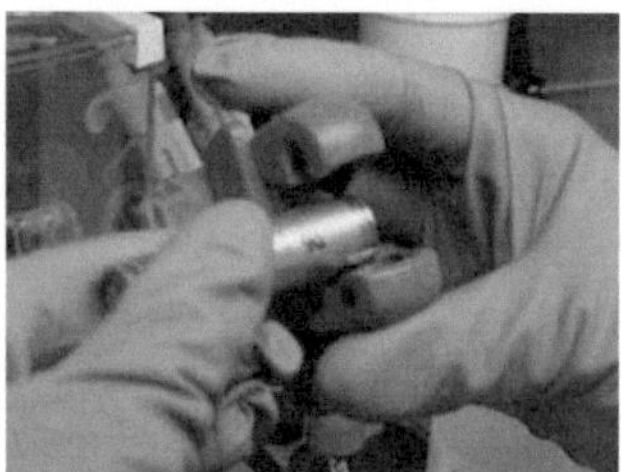

cut the top and the
bottom of the case with a
pipe-cutter

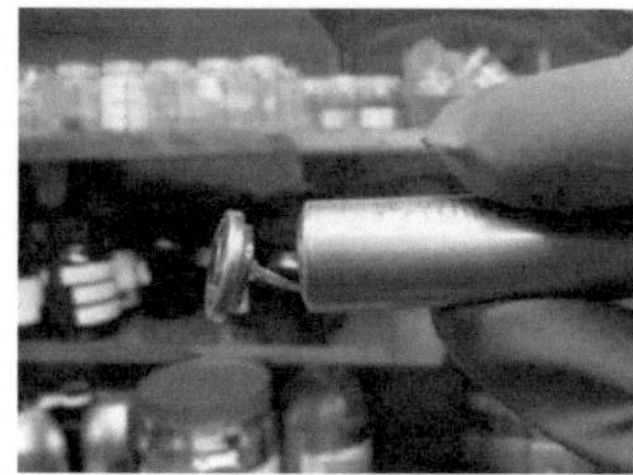

remove the top and the
bottom

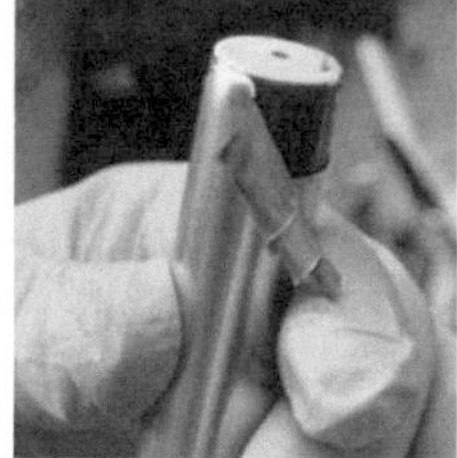

peel the cell case of
the electrode stack

$\longmapsto$

Only for double-side
coated electrodes

remove the plastic foil

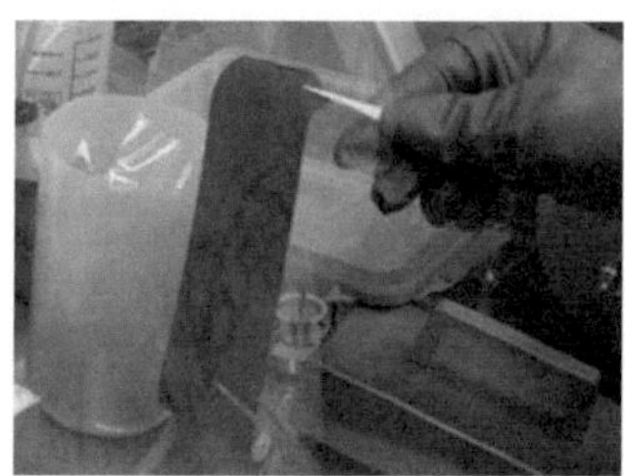

enroll and wash the
electrode stack

moisten the electrode with
an appropriate solvent

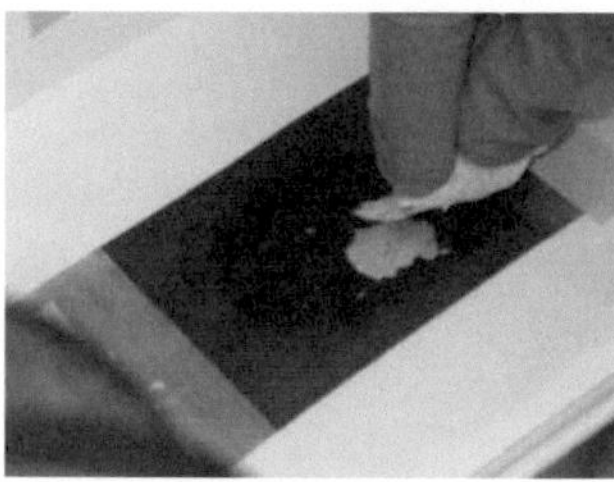

remove the electrode coating

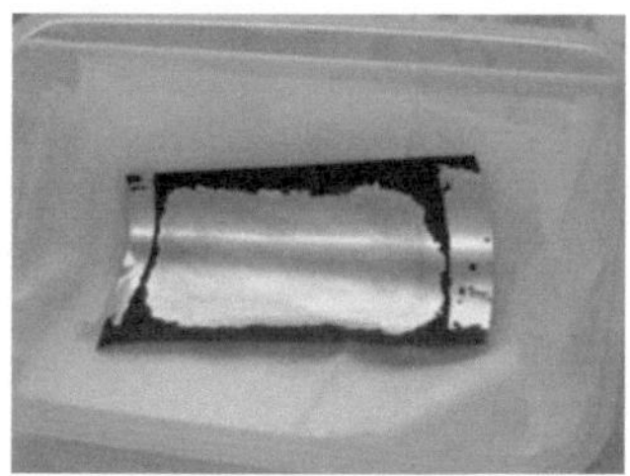

punch out and wash the
electrodes from the sheet
prepared

dry the electrodes
after preparation

C Measured Cells

cell ID	electrodes	measure-ments	reference electrode	comments
experimental cells				
AC02-0046	$LiFePO_4$/Li	EIS	-	temperature and SOC variation
AC02-0067	$2xLiFePO_4$	EIS	-	temperature variation
AC02-0119	Li/Li	EIS	-	temperature variation
AC02-0481	Li/Li	EIS	-	cycling
AC02-0232	Li/Li	EIS	copper wire	SOC variation
AC02-1336	$LiFePO_4$/Li	EIS	steel mesh	SOC variation
AC02-1334	$LiFePO_4$/Li	EIS	steel mesh	SOC variation
AC02-1182	$LiFePO_4$/Li	EIS	steel mesh	SOC variation
AC02-1180	$LiFePO_4$/Li	EIS	steel mesh	SOC variation
AC02-0574	$LiFePO_4$/Li	cycling	-	C-rate variation
AC02-1016	$LiFePO_4$/Li	cycling	-	C-rate variation
AC02-1039	$LiFePO_4$/Li	cycling	-	C-rate variation
AC02-1056	$LiFePO_4$/Li	cycling	-	C-rate variation
AJ02-1135	$LiFePO_4$/Li	EIS	steel mesh	SOC variation
AJ02-1797	$LiFePO_4$/Li	cycling	-	C-rate variation
AJ02-0437	$LiFePO_4$/Li	cycling	-	reproducibility
AJ02-0469	$LiFePO_4$/Li	cycling	-	reproducibility
AJ02-0495	$LiFePO_4$/Li	cycling	-	reproducibility
AJ02-0496	$LiFePO_4$/Li	cycling	-	reproducibility
AJ02-0497	$LiFePO_4$/Li	cycling	-	reproducibility
AJ02-0511	$LiFePO_4$/Li	cycling	-	reproducibility
AJ02-0512	$LiFePO_4$/Li	cycling	-	reproducibility
AJ02-0514	$LiFePO_4$/Li	cycling	-	reproducibility
AJ02-0526	$LiFePO_4$/Li	cycling	-	reproducibility
AJ02-0531	$LiFePO_4$/Li	cycling	-	reproducibility
AJ02-0536	$LiFePO_4$/Li	cycling	-	reproducibility
AJ02-0537	$LiFePO_4$/Li	cycling	-	reproducibility
AJ02-0539	$LiFePO_4$/Li	cycling	-	reproducibility
AJ02-0540	$LiFePO_4$/Li	cycling	-	reproducibility
AJ02-0543	$LiFePO_4$/Li	cycling	-	reproducibility
AJ02-0544	$LiFePO_4$/Li	cycling	-	reproducibility
AJ02-0548	$LiFePO_4$/Li	cycling	-	reproducibility
AJ02-0549	$LiFePO_4$/Li	cycling	-	reproducibility
AJ02-0555	$LiFePO_4$/Li	cycling	-	reproducibility
AJ02-0556	$LiFePO_4$/Li	cycling	-	reproducibility
AJ02-0557	$LiFePO_4$/Li	cycling	-	reproducibility

cell ID	electrodes	measurements	reference electrode	comments
experimental cells				
AJ02-0578	$LiFePO_4$/Li	cycling	-	reproducibility
AJ02-0579	$LiFePO_4$/Li	cycling	-	reproducibility
AJ02-0580	$LiFePO_4$/Li	cycling	-	reproducibility
AJ02-0581	$LiFePO_4$/Li	cycling	-	reproducibility
AJ02-0592	$LiFePO_4$/Li	cycling	-	reproducibility
AJ02-0593	$LiFePO_4$/Li	cycling	-	reproducibility
AJ02-0594	$LiFePO_4$/Li	cycling	-	reproducibility
AJ02-0595	$LiFePO_4$/Li	cycling	-	reproducibility
AJ02-0607	$LiFePO_4$/Li	cycling	-	reproducibility
AJ02-0608	$LiFePO_4$/Li	cycling	-	reproducibility
AJ02-0609	$LiFePO_4$/Li	cycling	-	reproducibility
AJ02-0613	$LiFePO_4$/Li	cycling	-	reproducibility
AJ02-0614	$LiFePO_4$/Li	cycling	-	reproducibility
AJ02-0625	$LiFePO_4$/Li	cycling	-	reproducibility
AJ02-0626	$LiFePO_4$/Li	cycling	-	reproducibility
AJ02-0627	$LiFePO_4$/Li	cycling	-	reproducibility
AJ02-0628	$LiFePO_4$/Li	cycling	-	reproducibility
AJ02-0676	$LiFePO_4$/Li	cycling	-	reproducibility
AJ02-0677	$LiFePO_4$/Li	cycling	-	reproducibility
AJ02-0678	$LiFePO_4$/Li	cycling	-	reproducibility
AJ02-0679	$LiFePO_4$/Li	cycling	-	reproducibility
AJ02-0680	$LiFePO_4$/Li	cycling	-	reproducibility
AJ02-0681	$LiFePO_4$/Li	cycling	-	reproducibility
AJ02-0682	$LiFePO_4$/Li	cycling	-	reproducibility
AJ02-0683	$LiFePO_4$/Li	cycling	-	reproducibility
AJ02-0829	C_6/Li	cycling	-	reproducibility
AJ02-0830	C_6/Li	cycling	-	reproducibility
AJ02-0831	C_6/Li	cycling	-	reproducibility
AJ02-0974	C_6/Li	cycling	-	reproducibility
AJ02-0975	C_6/Li	cycling	-	reproducibility
AJ02-1031	C_6/Li	cycling	-	reproducibility
AJ02-1032	C_6/Li	cycling	-	reproducibility
AJ02-0602	$LiFePO_4$/C_6	cycling	-	reproducibility
AJ02-0604	$LiFePO_4$/C_6	cycling	-	reproducibility
AJ02-0605	$LiFePO_4$/C_6	cycling	-	reproducibility
AJ02-0816	$LiFePO_4$/C_6	cycling	-	reproducibility
AJ02-0817	$LiFePO_4$/C_6	cycling	-	reproducibility

cell ID	electrodes	measurements	reference electrode	comments
experimental cells				
AJ02-0818	$LiFePO_4/C_6$	cycling	-	reproducibility
AJ02-0819	$LiFePO_4/C_6$	cycling	-	reproducibility
AJ02-0887	$LiFePO_4/C_6$	cycling	-	reproducibility
AJ02-0922	$LiFePO_4/C_6$	cycling	-	reproducibility
AJ02-0923	$LiFePO_4/C_6$	cycling	-	reproducibility
AJ02-0924	$LiFePO_4/C_6$	cycling	-	reproducibility
AJ02-0970	$LiFePO_4/C_6$	cycling	-	reproducibility
AJ02-0971	$LiFePO_4/C_6$	cycling	-	reproducibility
AJ02-1029	$LiFePO_4/C_6$	cycling	-	reproducibility
AJ02-1030	$LiFePO_4/C_6$	cycling	-	reproducibility
AJ02-1062	$LiFePO_4/C_6$	cycling	-	reproducibility
AJ02-1063	$LiFePO_4/C_6$	cycling	-	reproducibility
AJ02-1064	$LiFePO_4/C_6$	cycling	-	reproducibility
AJ02-1065	$LiFePO_4/C_6$	cycling	-	reproducibility
AJ02-1306	$LiFePO_4/C_6$	OCV	$Li_4Ti_5O_{12}$-mesh	-
AJ02-0970	$LiFePO_4/C_6$	TDM	-	-
AJ02-0973	$LiFePO_4/Li$	TDM	-	-
AJ02-0974	C_6/Li	TDM	-	-
AJ02-1376	C_6/Li	OCV	-	-
18650 cells				
AJ02-0786	$LiFePO_4/C_6$	EIS	-	temperature and SOC variation
AJ02-0968	$LiFePO_4/C_6$	EIS	-	temperature and SOC variation

D Supervised Diploma Theses and Study Projects

- Sabine Arnold, *Analyse des Impedanzspektrums von Lithium-Ionen Batterien mit Matlab*, study project 2010.

- Michael Weiss, *Modellierung der Festkörperdiffusion in Lithium-Ionen Batterien*, study project 2011.

- Kei Hirose, *Untersuchung von Be- und Entladeprozessen in Lithium-Ionen Batterie-Elektroden*, study project 2011.

- Simon Hansmann, *Modellierung und Simulation des Spannungsverhaltens einer Lithium-Ionen-Zelle*, (Co-supervised with A. Leonide), diploma thesis 2011.

- Andre Krefft, *Untersuchung von Lithium-Ionen Batterieelektroden aus geöffneten kommerziellen Zelle*, (Co-supervised with J. P. Schmidt), study project 2012.

- Tobias Talmon, *Untersuchung industriell gefertigter LiFePO4-Kathoden mit Impedanzmodellen*, study project 2012.

- Christian Uhlmann, *Abhängigkeit des Ladungstransferwiderstands vom Staging in Graphit*, (Co-supervised with M. Ender), seminar 2011.

- Amine Touzi, *Untersuchung eines 3-Elektrodensetups mittels Experimentalzellen und FEM-Simulationen*, (Co-supervised with M. Ender), study project 2012.

- Michael Weiss, *Untersuchung der niederfrequenten Impedanz von Lithium-Ionen Zellen durch Zeitbereichverfahren*, (Co-supervised with J. P. Schmidt), diploma thesis 2012.

- Wilko Sommer, *Aufbau und elektrochemische Charakterisierung von Halbzellen aus Anode und Kathode einer Li-Ionen Batterie*, external bachelor thesis 2011.

- Brad Bogdan, *Untersuchung des nichtlinearen Verhaltens einer Lithium-Ionen-Batterie*, (Co-supervised with J. P. Schmidt), study project 2013.

- Leander Teschner, *Beurteilung der Leistungsfähigkeit von Lithium-Ionen Batterie-Elektroden in Abhängigkeit von ihrer Mikrostruktur*, study project 2012.

- Christian Uhlmann, *Nachweis und Vorhersage von Lithium-Plating in Experimentalzellen*, (Co-supervised with M. Ender), diploma thesis 2012.

- Wilko Sommer, *Embedded Optical Sensors for Enhanced Management of Lithium-Ion Batteries*, external master thesis 2014.

E Publications

- J. Illig, J.P. Schmidt, M. Weiss, A. Weber and E. Ivers-Tiffée, *Understanding the impedance spectrum of 18650 LiFePO4-cells*, J. Power Sources. 239, p. 670-679 (2013)

- J. Illig, M. Ender, T. Chrobak, J. P. Schmidt, D. Klotz and E. Ivers-Tiffée, *Separation of Charge Transfer and Contact Resistance in LiFePO$_4$-Cathodes by Impedance Modeling*, J. Electrochem. Soc. 159, p. A952-A960 (2012).

- J. Illig, M. Ender, T. Chrobak, J. P. Schmidt, D. Klotz and E. Ivers-Tiffée, *Evaluation of LiFePO$_4$-Cathodes by Impedance Modeling*, in Dominique Guyomard (Ed.), Proceedings of the Lithium Batteries Discussion 2011, p. O11 (2011).

- J. Illig, T. Chrobak, D. Klotz and E. Ivers-Tiffée, *Evaluation of the Rate Determining Processes for LiFePO$_4$ as Cathode Material in Lithium-Ion-Batteries*, ECS Trans. 33, pp. 3-15 (2011).

- J. P. Schmidt, T. Chrobak, M. Ender, J. Illig, D. Klotz and E. Ivers-Tiffée, *Studies on LiFePO$_4$ as Cathode Material using Impedance Spectroscopy*, J. Power Sources 196, pp. 5342-5348 (2010).

- J. Illig, T. Chrobak, M. Ender, J. P. Schmidt, D. Klotz and E. Ivers-Tiffée, *Studies on LiFePO$_4$ as Cathode Material in Li-Ion Batteries*, ECS Trans. 28, pp. 3-17 (2010).

- T. Chrobak, M. Ender, J. Illig, J. P. Schmidt, D. Klotz and E. Ivers-Tiffée, *Studies on LiFePO$_4$ as Cathode Materials in Li-Ion-Batteries*, in DECHEMA (Ed.), First International Conference on Materials for Energy, Extended Abstracts - Book B: DECHEMA, pp. 593-595 (2010).

F Conference Contributions

- M. Weiss, J. Illig, J.P. Schmidt, M. Ender, A. Weber, E. Ivers-Tiffée, *Determination of Lithium Diffusion Coefficients in Commercial Electrodes by Impedance Modeling*, 224th ECS (San Francisco, USA), 28.10. - 1.11.2013

- J.P. Schmidt, J. Illig, A. Weber, E. Ivers-Tiffée, *Investigation of the μHz to mHz frequency range of commercial lithium-ion cells*, 224th ECS Meeting (San Francisco, USA), 27.10. - 01.11.2013

- J. Illig, J.P. Schmidt, M. Weiss, A. Weber, E. Ivers-Tiffée, *Multi-Step Approach for Impedance Analysis of 18650 Lithium-Ion Cells*, 9th International Symposium on Electrochemical Impedance Spectroscopy (Okinawa, Japan), 17.06. - 21.06.2013

- C. Uhlmann, J. Illig, M. Ender, A. Weber, R. Schuster, E. Ivers-Tiffée, *In situ detection of lithium metal plating on graphite via reference electrodes and optical test-cells*, 223rd Meeting of The Electrochemical Society (Toronto, Canada), 12.05. - 17.05.2013

- J.P. Schmidt, J. Illig, A. Weber, E. Ivers-Tiffée, *Lifetime-Tests for Lithium-Ion Cells via Current Pulse Measurements*, 2nd International Conference on Materials for Energy, EnMat II (Karlsruhe), 12.06. - 16.06.2013

- J. Illig, M. Ender, J.P. Schmidt, A. Weber, E. Ivers-Tiffée, *Combination of Impedance Modelling and Microstructure Analysis for LiFePO$_4$-Cathodes*, 2nd International Conference on Materials for Energy, EnMat II (Karlsruhe), 12.06. - 16.06.2013

- J. Illig, J.P. Schmidt, M. Weiss, A. Weber, E. Ivers-Tiffée, *Understanding the impedance spectrum of 18650 LiFePO$_4$-cells*, Kraftwerk Batterie 2013 (Aachen), 25.02. - 27.02.2013

- J. Illig, M. Ender, A. Weber, E. Ivers-Tiffée, *Reference Electrodes for Impedance Measurements in Lithium Ion Cells*, 222nd ECS-Meeting (Honolulu, Hawaii), 11.10. - 11.10.2012

- J. Illig, M. Ender, A. Weber, E. Ivers-Tiffée, *Comparison of Charge Transfer Resistance for different $LiFePO_4$-Electrodes*, 222nd ECS-Meeting (Hawaii, Honolulu), 08.10. - 2012

- M. Ender, J. Illig, E. Ivers-Tiffée, *Parameterizing Li-Ion Cell Models Supported by Microstructure Reconstructions*, 222nd Meeting of The Electrochemical Society (Honolulu, USA), 07.10. - 12.10.2012

- J. Illig, J.P. Schmidt, M. Weiss, A. Weber, E. Ivers-Tiffée, *Understanding the impedance of 18650 cells by experimental cell measurements*, IWIS (Chemnitz, Germany), 27.09. - 27.09.2012

- J. Illig, J.P. Schmidt, M. Weiss, E. Ivers-Tiffée, *Quantification of Impedance Contributions of $LiFePO_4$ - and Graphite-Electrode to the Total Impedance of a Commercial 18650 Lithium-Ion Cell*, UECT 2012 (Ulm), 03.07. - 05.07.2012

- J.P. Schmidt, M. Schönleber, D. Klotz, J. Illig, A. Weber, E. Ivers-Tiffée, *Investigation of the μHz-Range in Impedance Spectra of Li-Ion Cells*, 16th International Meeting on Lithium Batteries (Jeju, Korea), 17.06. - 22.06.2012

- J. Illig, J.P. Schmidt, M. Weiss, A. Weber, E. Ivers-Tiffée, *Contributions of $LiFePO_4$- and Graphite-Electrode to the Total Impedance of a 18650 Lithium-Ion Cell*, 16th IMLB (Jeju, Island), 18.06. - 18.06.2012

- J. Illig, M. Ender, E. Ivers-Tiffée, *Evaluation of Commercial $LiFePO_4$ Cathodes by Impedance Modeling*, 220th Meeting of The Electrochemical Society (Boston, USA), 11.10. - 15.10.2011

- J. Illig, S. Hansmann, A. Leonide, E. Ivers-Tiffée, *Detailed Electrochemical Analysis and Modeling of NMC-Based Lithium Ion Cells*, 220th Meeting of The Electrochemical Society (Boston, USA), 11.10. - 15.10.2011

- J. Illig, M. Ender, T. Chrobak, J.P. Schmidt, D. Klotz, E. Ivers-Tiffée, *Evaluation of $LiFePO_4$-Cathodes by Impedance Modeling*, Lithium Batteries Discussion 2011 (Arcachon, France), 12.06. - 17.06.2011

- J. Illig, T. Chrobak, D. Klotz, E. Ivers-Tiffée, *Evaluation of the Rate Determining Processes for $LiFePO_4$ as Cathode Material in Lithium-Ion-Batteries*, 218th Meeting of The Electrochemical Society (Las Vegas, USA), 10.10. - 15.10.2010

- M. Ender, T. Chrobak, J. Illig, J.P. Schmidt, D. Klotz, E. Ivers-Tiffée, *Identification of Reaction Mechanisms in Lithium-Ion Cells by Deconvolution of Electrochemical Impedance Spectra*, International Meeting on Lithium Batteries (Montreal, Canada), 27.06. - 02.07.2010

- J.P. Schmidt, T. Chrobak, M. Ender, J. Illig, D. Klotz, E. Ivers-Tiffée, *Studies on LiFePO$_4$ as Cathode Material using Impedance Spectroscopy*, 12th Ulm Electro-Chemical Talks (UECT) (Ulm, Germany), 16.06. - 17.06.2010

- T. Chrobak, M. Ender, J. Illig, J.P. Schmidt, D. Klotz, E. Ivers-Tiffée, *Studies on LiFePO$_4$ as Cathode Material in Li-Ion-Batteries*, First International Conference on Materials for Energy (Karlsruhe, Germany), 05.07. - 08.07.2010

- J. Illig, T. Chrobak, M. Ender, J.P. Schmidt, D. Klotz, E. Ivers-Tiffée, *Studies on LiFePO$_4$ as Cathode Material in Li-Ion Batteries*, 217th Meeting of The Electrochemical Society (Vancouver, Canada), 26.04. - 30.04.2010

Acronyms

aASR	active Area Specific Resistance, using the active surface area between active material and electrolyte, previously determined by microstructure reconstruction, for normalization
ASR	Area Specific Resistance
CB	Carbon Black
CC	Constant Current
CCCV	Constant Current Constant Voltage
CV	Constant Voltage
CNLS	Complex Non-linear Least Squares Fit
CPE	Constant Phase Element
CT	Charge Transfer
DRT	Distribution of Relaxation Times
DEC	Diethyl Carbonate
DMC	Dimethyl Carbonate
EC	Ethylene Carbonate
ECM	Equivalent Circuit Model
EIS	Electrochemical Impedance Spectroscopy
EMC	Ethyl Methyl Carbonate
EV	Electric Vehicle
FEM	Finite Element Method
FIB	Focused Ion Beam
FLW	Finite Length Warburg Element
FRA	Frequency Response Analyzer
FSW	Finite Space Warburg Element
FTIR	Fourier Transform Infrared Spectroscopy
GFLW	Generalized Finite Length Warburg Element
GFSW	Generalized Finite Space Warburg Element
GITT	Galvanostatic Intermittent Titration Technique
HCl	Hydrochloric Acid
HN	Havriliak Negami
HOPG	Highly Oriented Pyrolytic Graphite
ISC	Fraunhofer Institut für Silicatforschung
IWE	Institut für Werkstoffe der Elektrotechnik
KK	Kramers Kronig
MCMB	Meso Carbon Micro Beads
MWCNT	multi-walled carbon nanotubes

NCA	$LiNi_xCo_yAl_{1-x-y}$
NMC	$LiNi_xMn_yCo_{1-x-y}$
NMP	N-methyl-2-pyrrolidone
OCP	Open Circuit Potential
OCV	Open Circuit Voltage
PC	Propylene Carbonate
PE	Polyethylene
PITT	Potentiostatic Intermittent Titration Technique
PP	Polypropylene
PVDF	Poly (Vinylidene Fluoride)-Binder
RMS	Root Mean Square
RMSE	Root Mean Squares Error
RVE	Representative Volume Element
SEI	Solid Electrolyte Interphase
SEM	Scanning Electron Microscope
SOC	State Of Charge
SOFC	Solid Oxide Fuel Cell
SOH	State Of Health
TDM	Time Domain Measurements
TLM	Transmission Line Model
VC	Vinylene Carbonate
VGCF	Vapor Grown Nano Fibers
XRD	X-ray Diffraction

Symbols

Symbols

A	electrode area
C_{cell}	cell capacity
$C_{spec,M}$	mass specific capacity
$C_{spec,V}$	volume specific capacity
$C_{\Delta int}$	intercalation capacity
E_{act}	activation energy
D	diffusion coefficient
F	Farraday constant
ΔG	Gibbs free energy
I_0/I_1	Bessel functions of the first kind
L	diffusion length
M_{Molar}	molar mass
Q	charge
R_g	universal gas constant
U_{cell}	cell voltage
U_{max}	maximum cell voltage
U_{min}	minimum cell voltage
U_{OCV}	open circuit voltage
U_{th}	theoretical cell voltage
V_{Molar}	molar volume
Z	complex impedance
Z'/Z''	real part and imaginary part of the complex impedance

Symbols

$a_{LFP}, a_{CB}, a_{porosity}, a_{active}$	specific surface areas
c	concentration
c_0	equilibrium concentration
f_{char}	characteristic frequency
f_s	sample rate
m_{cell}	cell mass
n	number of charge carriers
res_{real}	real part of fit residuals
res_{imag}	imaginary part of fit residuals
$w_{th,M}$	mass specific theoretical energy density
$w_{th,V}$	volume specific theoretical energy density
$w_{pr,M}$	practical mass specific energy density
$w_{pr,V}$	practical volume specific energy density
ϵ	porosity
γ/g	continuous/discrete distribution of relaxation times
λ	regularization parameter
ω	angular frequency
ϕ	phase angle
$\sigma_{electron}$	electronic conductivity
σ_{ion}	specific conductivity
$\sigma_{ion,eff}$	effective specific conductivity
τ	time constant
τ_p	tortuosity
τ_{min}	minimum time constant
τ_{max}	maximum time constant
τ_{relax}	relaxation time constant

Bibliography

[1] Moses Ender. *Mikrostrukturelle Charakterisierung, Modellentwicklung und Simulation poröser Elektroden fuer Lithiumionenzellen*. KIT Scientific Publishing, 2014.

[2] Florian Hannig, Tom Smolinka, Peter Bretschneider, Steffen Nicolai, Sven Krüger, Frank Meißner, and Marco Voigt. Stand und Entwicklungspotenzial der Speichertechniken für Elektroenergie–Ableitung von Anforderungen an und Auswirkungen auf die Investitionsgüterindustrie. *Abschlussbericht der BMWi-Auftragsstudie*, 8:28, 2009.

[3] David Linden and Thomas B Reddy. *Handbook of batteries*. McGraw-Hill, 2002. third edition.

[4] Hai-Yen Tran. *Verfahrenstechnische Entwicklung und Untersuchung von Elektroden und deren Herstellprozess für innovative Lithium-Hochleistungsbatterien*. PhD thesis, ZSW, 2011.

[5] Hai Yen Tran, Corina Täubert, Meike Fleischhammer, Peter Axmann, Laszlo Küppers, and Margret Wohlfahrt-Mehrens. $LiMn_2O_4$ spinel/$LiNi_{0,8}Co_{0,15}Al_{0,05}O_2$ blends as cathode materials for lithium-ion batteries. *Journal of The Electrochemical Society*, 158(5):A556–A561, 2011.

[6] Martin Winter, Jürgen O. Besenhard, Michael E. Spahr, and Petr Novák. Insertion electrode materials for rechargeable lithium batteries. *Advanced Materials*, 10(10):725–763, 1998.

[7] Simon Theil, Meike Fleischhammer, Peter Axmann, and Margret Wohlfahrt-Mehrens. Experimental investigations on the electrochemical and thermal behaviour of $LiCoPO_4$-based cathode. *Journal of Power Sources*, 222(0):72 – 78, 2013.

[8] Claus Daniel and Jürgen O. Besenhard. *Handbook of Battery Materials*. Wiley-VCH, 2011.

[9] Andreas Jossen, Wolfgang Weydanz, et al. *Moderne Akkumulatoren richtig einsetzen*. Inge Reichardt Verlag, 2006.

[10] Robert Huggins. *Advanced batteries: materials science aspects*. Springer, 2009.

[11] K. Mizushima, P.C. Jones, P.J. Wiseman, and J.B. Goodenough. Li_xCoO_2: A new cathode material for batteries of high energy density. *Materials Research Bulletin*, 15(6):783 – 789, 1980.

[12] Zhaolin Liu, Aishui Yu, and Jim Lee. Synthesis and characterization of NCM as the cathode materials of secondary lithium batteries. *Journal of Power Sources*, 81-82(0):416–419, 1999.

[13] Masaki Yoshio, Hideyuki Noguchi, Jun ichi Itoh, Masaki Okada, and Takashi Mouri. Preparation and properties of NMC as a cathode for lithium ion batteries. *Journal of Power Sources*, 90(2):176 – 181, 2000.

[14] M. Stanley Whittingham. Lithium batteries and cathode materials. *Chemical Reviews*, 104(10):4271–4302, 2004.

[15] Jeffrey W. Fergus. Recent developments in cathode materials for lithium ion batteries. *Journal of Power Sources*, 195(4):939 – 954, 2010.

[16] J. Weaving, F. Coowar, D. Teagle, J. Cullen, V. Dass, P. Bindin, R. Green, and W. Macklin. Development of high energy density li-ion batteries based on NCA. *Journal of Power Sources*, 97-98:733–735, 2001.

[17] S. Albrecht, J. Kuempers, M. Kruft, S. Malcus, C. Vogler, M. Wahl, and M. Wohlfahrt-Mehrens. Electrochemical and thermal behavior of aluminum- and magnesium-doped spherical lithium nickel cobalt mixed oxides $Li_{1-x}(Ni_{1-y-z}Co_yM_z)O_2$ (M=Al,Mg). *Journal of Power Sources*, 119(0):178–183, 2003.

[18] Chang Joo Han, Jang Hyuk Yoon, Won Il Cho, and Ho Jang. Electrochemical properties of $LiNi_{0.8}Co_{0.2-x}Al_xO_2$ prepared by a sol-gel method. *Journal of Power Sources*, 136(1):132 – 138, 2004.

[19] Sun-Ho Kang, Won-Sub Yoon, Kyung-Wan Nam, Xiao-Qing Yang, and Daniel P. Abraham. Investigating the first-cycle irreversibility of lithium metal oxide cathodes for li batteries. *Journal of Materials Science*, 43(14):4701–4706, 2008.

[20] Hyun Joo Bang, Humberto Joachin, Hui Yang, Khalil Amine, and Jai Prakash. Contribution of the structural changes of $LiNi_{0.8}Co_{0.15}Al_{0.05}O_2$ cathodes on the exothermic reactions in li-ion cells. *Journal of The Electrochemical Society*, 153(4):A731–A737, 2006.

[21] Tsutomu Ohzuku, Masaki Kitagawa, and Taketsugu Hirai. Electrochemistry of manganese dioxide in lithium nonaqueous cell III. *Journal of The Electrochemical Society*, 137(3):769–775, 1990.

[22] R.J. Gummow, A. de Kock, and M.M. Thackeray. Improved capacity retention in rechargeable 4V lithium/lithium-manganese oxide (spinel) cells. *Solid State Ionics*, 69(1):59 – 67, 1994.

[23] Mamoru Hosoya, Hiromasa Ikuta, Takashi Uchida, and Masataka Wakihara. The defect structure model in nonstoichiometric $LiMn_2O_4$. *Journal of The Electrochemical Society*, 144(4):L52–L53, 1997.

[24] A. K. Padhi, K. S. Nanjundaswamy, and J. B. Goodenough. Phospho-olivines as positive-electrode materials for rechargeable lithium batteries. *Journal of The Electrochemical Society*, 144(4):1188–1194, 1997.

[25] John B Goodenough, Akshaya K Padhi, KS Nanjundaswamy, and Christian Masquelier. Cathode materials for secondary (rechargeable) lithium batteries, June 8 1999. US Patent 5,910,382.

[26] Arumugam Manthiram and Theivanayagam Muraliganth. *In: Handbook of Battery Materials, Editor: Claus Daniel and Juergen O. Besenhard.* Wiley-VCH, 2011.

[27] J.M. Tarascon, C. Delacourt, A.S. Prakash, M. Morcrette, M.S. Hegde, C. Wurm, and C. Masquelier. Various strategies to tune the ionic/electronic properties of electrode materials. *Dalton Transactions*, 19:2988–2994, 2004.

[28] Daniela Zane, Maria Carewska, Silvera Scaccia, Francesco Cardellini, and Pier Paolo Prosini. Factor affecting rate performance of undoped $LiFePO_4$. *Electrochimica Acta*, 49(25):4259–4271, 2004.

[29] A. Vadivel Murugan, T. Muraliganth, and A. Manthiram. Rapid microwave-solvothermal synthesis of phospho-olivine nanorods and their coating with a mixed conducting polymer for lithium ion batteries. *Electrochemistry Communications*, 10(6):903 – 906, 2008.

[30] Karim Zaghib, Chiaki Sotowa, Patrick Charest, Masataka Takeuchi, and Abdelbast Guerfi. Composite electrode material, 2009. US Patent App. 12/867,414.

[31] Venkat Srinivasan and John Newman. Discharge model for the lithium iron-phosphate electrode. *Journal of The Electrochemical Society*, 151(10):A1517–A1529, 2004.

[32] Venkat Srinivasan and John Newman. Existence of path-dependence in the $LiFePO_4$ electrode. *Electrochemical and solid-state letters*, 9(3):A110–A114, 2006.

[33] William C. Chueh, Farid El Gabaly, Joshua D. Sugar, Norman C. Bartelt, Anthony H. McDaniel, Kyle R. Fenton, Kevin R. Zavadil, Tolek Tyliszczak, Wei Lai, and Kevin F. McCarty. Intercalation pathway in many-particle $LiFePO_4$ electrode revealed by nanoscale state-of-charge mapping. *Nano Letters*, 13(3):866–872, 2013.

[34] Jun ichi Yamaki and Shin ichi Tobishima. *In: Handbook of Battery Materials, Editor: Claus Daniel and Juergen O. Besenhard.* Wiley-VCH, 2011.

[35] Robert A. Huggins. *In: Handbook of Battery Materials, Editor: Claus Daniel and Juergen O. Besenhard.* Wiley-VCH, 2011.

[36] Kazunori Ozawa. Llithium-ion rechargeable batteries with $LiCoO_2$ and carbon electrodes: the $LiCoO_2C$ system. *Solid State Ionics*, 69(3):212 – 221, 1994.

[37] Martin Winter and Juergen Otto Besenhard. *In: Handbook of Battery Materials, Editor: Claus Daniel and Juergen O. Besenhard.* Wiley-VCH, 2011.

[38] Rosalind E. Franklin. Crystallite growth in graphitizing and non-graphitizing carbons. *Proceedings of the Royal Society of London. Series A. Mathematical and Physical Sciences*, 209(1097):196–218, 1951.

[39] Jürgen O. Besenhard and Heinz P. Fritz. The electrochemistry of black carbons. *Angewandte Chemie International Edition in English*, 22(12):950–975, 1983.

[40] Nobuyuki Imanishi, Yasuo Takeda, and Osamu Yamamoto. *Development of the Carbon Anode in Lithium Ion Batteries*, pages 98–126. Wiley-VCH Verlag GmbH, 2007.

[41] Diane Golodnitsky Emanuel Peled and Jack Penciner. *In: Handbook of Battery Materials, Editor: Claus Daniel and Juergen O. Besenhard.* Wiley-VCH, 2011.

[42] Kang Xu. Nonaqueous liquid electrolytes for lithium-based rechargeable batteries. *Chemical reviews*, 104(10):4303–4418, 2004.

[43] Kang Xu. Charge-transfer process at graphite electrolyte interface and the solvation sheath structure of li in nonaqueous electrolytes. *Journal of The Electrochemical Society*, 154(3):A162–A167, 2007.

[44] Kang Xu, Yiufai Lam, Sheng S. Zhang, T. Richard Jow, and Timothy B. Curtis. Solvation sheath of li^+ in nonaqueous electrolytes and its implication of graphite/electrolyte interface chemistry. *The Journal of Physical Chemistry C*, 111(20):7411–7421, 2007.

[45] Selectilyte Merck/BASF. Datasheet for Selectilyte Electrolyte Series.

[46] D. Aurbach, K. Gamolsky, B. Markovsky, Y. Gofer, M. Schmidt, and U. Heider. On the use of vinylene carbonate (VC) as an additive to electrolyte solutions for li-ion batteries. *Electrochimica Acta*, 47(9):1423 – 1439, 2002.

[47] Sheng Shui Zhang. A review on electrolyte additives for lithium-ion batteries. *Journal of Power Sources*, 162(2):1379 – 1394, 2006.

[48] E. Peled, D. Golodnitsky, G. Ardel, and V. Eshkenazy. The SEI model-application to lithium-polymer electrolyte batteries. *Electrochimica Acta*, 40(13):2197 – 2204, 1995.

[49] Pankaj Arora, Ralph E. White, and Marc Doyle. Capacity fade mechanisms and side reactions in lithium-ion batteries. *Journal of The Electrochemical Society*, 145(10):3647–3667, 1998.

[50] Matthew B. Pinson and Martin Z. Bazant. Theory of sei formation in rechargeable batteries: Capacity fade, accelerated aging and lifetime prediction. *Journal of The Electrochemical Society*, 160(2):A243–A250, 2013.

[51] Andreas Jossen. Fundamentals of battery dynamics. *Journal of Power Sources*, 154(2):530 – 538, 2006.

[52] Lihua Shen and Zhangxin Chen. Critical review of the impact of tortuosity on diffusion. *Chemical Engineering Science*, 62(14):3748 – 3755, 2007.

[53] Evgenij Barsoukov and J Ross Macdonald. *Impedance spectroscopy: theory, experiment, and applications.* Wiley.com, 2005.

[54] Torben Jacobsen and Keld West. Diffusion impedance in planar, cylindrical and spherical symmetry. *Electrochimica Acta*, 40(2):255 – 262, 1995.

[55] Ellen Ivers-Tiffée. Batterien und Brennstoffzellen WS 12-13. University Lecture, 2012.

[56] Kohei Honkura, Hidetoshi Honbo, Yoshimasa Koishikawa, and Tatsuo Horiba. State analysis of lithium-ion batteries using discharge curves. *ECS Transactions*, 13(19):61–73, 2008.

[57] David Linden and Thomas B Reddy. Handbook of batteries, third edition. *New York*, 2002.

[58] Mark E Orazem and Bernard Tribollet. *Electrochemical Impedance Spectroscopy.* Wiley VHC, 2008.

[59] M. D. Levi and D. Aurbach. Simultaneous measurements and modeling of the electrochemical impedance and the cyclic voltammetric characteristics of graphite electrodes doped with lithium. *The Journal of Physical Chemistry B*, 101(23):4630–4640, 1997.

[60] Jörg Illig, Jan Philipp Schmidt, Michael Weiss, André Weber, and Ellen Ivers-Tiffée. Understanding the impedance spectrum of 18650 $LiFePO_4$-cells. *Journal of Power Sources*, 239(0):670 – 679, 2013.

[61] Uwe Kiencke and Holger Jaekel. *Signale und Systeme.* Oldenbourg Verlag, 2008.

[62] D. Klotz, M. Schoenleber, J.P. Schmidt, and E. Ivers-Tiffée. New approach for the calculation of impedance spectra out of time domain data. *Electrochimica Acta*, 56(24):8763 – 8769, 2011.

[63] Bernard A. Boukamp and J. Ross Macdonald. Alternatives to kronig-kramers transformation and testing, and estimation of distributions. *Solid State Ionics*, 74(1):85–101, 1994.

[64] Bernard A. Boukamp. A linear kronig-kramers transform test for immittance data validation. *Journal of The Electrochemical Society*, 142(6):1885–1894, 1995.

[65] Pankaj Agarwal, Mark E. Orazem, and Luis H. Garcia-Rubio. Measurement models for electrochemical impedance spectroscopy: I. demonstration of applicability. *Journal of The Electrochemical Society*, 139(7):1917–1927, 1992.

[66] Egon Ritter v. Schweidler. Studien über die Anomalien im Verhalten der Dielektrika. *Annalen der Physik*, 329(14):711–770, 1907.

[67] Karl Willy Wagner. Zur Theorie der unvollkommenen Dielektrika. *Annalen der Physik*, 345(5):817–855, 1913.

[68] H. Schichlein, A.C. Müller, M. Voigts, A. Krügel, and E. Ivers-Tiffée. Deconvolution of electrochemical impedance spectra for the identification of electrode reaction mechanisms in solid oxide fuel cells. *Journal of Applied Electrochemistry*, 32(8):875–882, 2002.

[69] Helge Schichlein. *Experimentelle Modellbildung für die Hochtemperaturbrennstoffzelle SOFC*. Universität Karlsruhe, 2003.

[70] A. Leonide, V. Sonn, A. Weber, and E. Ivers-Tiffée. Evaluation and modeling of the cell resistance in anode-supported solid oxide fuel cells. *Journal of The Electrochemical Society*, 155(1):B36–B41, 2008.

[71] Andre Leonide. *SOFC Modelling and Parameter Identification by means of Impedance Spectroscopy*. KIT Scientific Publishing, 2010.

[72] V. Sonn, A. Leonide, and E. Ivers-Tiffée. Combined deconvolution and CNLS fitting approach applied on the impedance response of technical Ni 8YSZ cermet electrodes. *Journal of The Electrochemical Society*, 155(7):B675–B679, 2008.

[73] Andrei Nikolaevich Tikhonov. *Numerical methods for the solution of ill-posed problems*. Kluwer Academic Publishers Dordrecht, 1995.

[74] Jan Philipp Schmidt. *Verfahren zur Charakterisierung und Modellierung von Lithium-Ionen Zellen*. KIT Scientific Publishing, 2013.

[75] PNGV battery test manual, 2001.

[76] Electrically propelled road vehicles - test specification for lithium-ion traction battery systems - part 1: High energy applications, 2009.

[77] W. Weppner and R.A. Huggins. Electrochemical investigation of the chemical diffusion, partial ionic conductivities, and other kinetic parameters in Li_3Sb and Li_3Bi. *Journal of Solid State Chemistry*, 22(3):297 – 308, 1977.

[78] C. John Wen, B. A. Boukamp, R. A. Huggins, and W. Weppner. Thermodynamic and mass transport properties of LiAl. *Journal of The Electrochemical Society*, 126(12):2258–2266, 1979.

[79] Dennis W. Dees, Shigehiro Kawauchi, Daniel P. Abraham, and Jai Prakash. Analysis of the galvanostatic intermittent titration technique (GITT) as applied to a lithium-ion porous electrode. *Journal of Power Sources*, 189(1):263 – 268, 2009.

[80] Jan Philipp Schmidt, Hai Yen Tran, Jan Richter, Ellen Ivers-Tiffée, and Margret Wohlfahrt-Mehrens. Analysis and prediction of the open circuit potential of lithium-ion cells. *Journal of Power Sources*, 239(0):696 – 704, 2013.

[81] Rachid Yazami and Yvan F. Reynier. Mechanism of self-discharge in graphite lithium anode. *Electrochimica Acta*, 47(8):1217 – 1223, 2002.

[82] N. N. Sinha, A. J. Smith, J. C. Burns, Gaurav Jain, K. W. Eberman, Erik Scott, J. P. Gardner, and J. R. Dahn. The use of elevated temperature storage experiments to learn about parasitic reactions in wound $LiCoO_2$/graphite cells. *Journal of The Electrochemical Society*, 158(11):A1194–A1201, 2011.

[83] Chunsheng Wang, Xiang wu Zhang, A.John Appleby, Xiaole Chen, and Frank E Little. Self-discharge of secondary lithium-ion graphite anodes. *Journal of Power Sources*, 112(1):98 – 104, 2002.

[84] J. Ross Macdonald. Note on the parameterization of the constant-phase admittance element. *Solid State Ionics*, 13(2):147 – 149, 1984.

[85] M.R. Shoar Abouzari, F. Berkemeier, G. Schmitz, and D. Wilmer. On the physical interpretation of constant phase elements. *Solid State Ionics*, 180(14):922 – 927, 2009.

[86] J. Bisquert, G. Garcia-Belmonte, P. Bueno, E. Longo, and L.O.S. Bulhoes. Impedance of constant phase element (CPE)-blocked diffusion in film electrodes. *Journal of Electroanalytical Chemistry*, 452(2):229 – 234, 1998.

[87] Bryan Hirschorn, Mark E. Orazem, Bernard Tribollet, Vincent Vivier, Isabelle Frateur, and Marco Musiani. Determination of effective capacitance and film thickness from constant-phase-element parameters. *Electrochimica Acta*, 55(21):6218 – 6227, 2010.

[88] J.Ross Macdonald. New aspects of some small-signal ac frequency response functions. *Solid State Ionics*, 15(2):159 – 161, 1985.

[89] Kenneth S. Cole and Robert H. Cole. Dispersion and absorption in dielectrics I. alternating current characteristics. *The Journal of Chemical Physics*, 9(4):341–351, 1941.

[90] Dino Klotz. *Characterization and Modeling of Electrochemical Energy Conversion Systems by Impedance Techniques*. KIT Scientific Publishing, 2012.

[91] S. Havriliak and S. Negami. A complex plane representation of dielectric and mechanical relaxation processes in some polymers. *Polymer*, 8(0):161 – 210, 1967.

[92] E. Sanchez Martinez, R. Diaz Calleja, and W. Gunsser. Complex polarizability as used to analyze dielectric relaxation measurements. *Colloid and Polymer Science*, 270(2):146–153, 1992.

[93] N.K. Karan, O.K. Pradhan, R. Thomas, B. Natesan, and R.S. Katiyar. Solid polymer electrolytes based on polyethylene oxide and lithium trifluoro- methane sulfonate (PEO-$LiCF_3SO_3$): Ionic conductivity and dielectric relaxation. *Solid State Ionics*, 179(19-20):689–696, 2008.

[94] Bernard A. Boukamp and Henny J.M. Bouwmeester. Interpretation of the gerischer impedance in solid state ionics. *Solid State Ionics*, 157(1):29–33, 2003. Proceedings of the 6th International Symposium on Systems with Fast Ionic Transport (ISSFIT).

[95] Juan Bisquert. Theory of the impedance of electron diffusion and recombination in a thin layer. *The Journal of Physical Chemistry B*, 106(2):325–333, 2002.

[96] Bernard A. Boukamp. Electrochemical impedance spectroscopy in solid state ionics: recent advances. *Solid State Ionics*, 169(1):65–73, 2004. Proceedings of the Annual Meeting of International Society of Electrochemistry.

[97] D. Johnson, 2002. ZView2 Help.

[98] Milton Abramowitz and Irene A. Stegun. *Handbook of Mathematical Functions: With Formulars, Graphs, and Mathematical Tables*, volume 55. DoverPublications.com, 1964.

[99] J.E.B. Randles. Kinetics of rapid electrode reactions - part 1. *Discussions of the faraday society*, 1:11–19, 1947.

[100] J.E.B. Randles. Kinetics of rapid electrode reactions part 2 - rate constants and activation energies of electrode reactions. *Transactions of the Faraday Society*, 48:828–832, 1952.

[101] J.E.B. Randles and K.W. Somertron. Kinetics of rapid electrode reactions, part 3 - electron exchange reactions. *Transactions of the Faraday Society*, 48:937–950, 1952.

[102] Evgenij Barsoukov, Jong Hyun Kim, Jong Hun Kim, Chul Oh Yoon, and Hosull Lee. Kinetics of lithium intercalation into carbon anodes in situ impedance investigation of thickness and potential dependence. *Solid State Ionics*, 116(3-4):249–261, 1999.

[103] Moses Ender, André Weber, and Ellen Ivers-Tiffée. A novel method for measuring the effective conductivity and the contact resistance of porous electrodes for lithium-ion batteries. *Electrochemistry Communications*, 34(0):130 – 133, 2013.

[104] J. Euler and W. Nonnenmacher. Stromverteilung in poroesen Elektroden. *Electrochimica Acta*, 2(4):268 – 286, 1960.

[105] R. de Levie. On porous electrodes in electrolyte solutions: I. capacitance effects. *Electrochimica Acta*, 8(10):751 – 780, 1963.

[106] J. H. Sluyters and J. J. C. Oomen. On the impedance of galvanic cells: II. experimental verification. *Recueil des Travaux Chimiques des Pays-Bas*, 79(10):1101–1110, 1960.

[107] James Ross MacDonald. Impedence spectroscopy–emphasizing solid materials and systems. *Wiley-Interscience, John Wiley and Sons,*, pages 1–346, 1987.

[108] Masaya Takahashi, Shin ichi Tobishima, Koji Takei, and Yoji Sakurai. Reaction behavior of $LiFePO_4$ as a cathode material for rechargeable lithium batteries. *Solid State Ionics*, 148(3):283 – 289, 2002. Proceedings of the Symposium on Materials for Advanced Batteries and Fuel Cells. Organised in conjunction with

the International Conference on Materials for Advanced Technologies (ICMAT 2001).

[109] S. Franger, F. Le Cras, C. Bourbon, and H. Rouault. $LiFePO_4$ synthesis routes for enhanced electrochemical performance. *Electrochemical and Solid-State Letters*, 5(10):A231–A233, 2002.

[110] Pier Paolo Prosini, Marida Lisi, Daniela Zane, and Mauro Pasquali. Determination of the chemical diffusion coefficient of lithium in $LiFePO_4$. *Solid State Ionics*, 148(1):45 – 51, 2002.

[111] Atsushi Funabiki, Minoru Inaba, Zempachi Ogumi, Shin-ichi Yuasa, Junhiko Otsuji, and Akimasa Tasaka. Impedance study on the electrochemical lithium intercalation into natural graphite powder. *Journal of The Electrochemical Society*, 145(1):172–178, 1998.

[112] Y.C. Chang and H.J. Sohn. Electrochemical impedance analysis for lithium ion intercalation into graphitized carbons. *Journal of the Electrochemical Society*, 147(1):50–58, Jan 2000.

[113] Fabio La Mantia, Jens Vetter, and Petr Novák. Impedance spectroscopy on porous materials: A general model and application to graphite electrodes of lithium-ion batteries. *Electrochimica Acta*, 53(12):4109 – 4121, 2008.

[114] M. Umeda, K. Dokko, Y. Fujita, M. Mohamedi, I. Uchida, and J.R. Selman. Electrochemical impedance study of li-ion insertion into mesocarbon microbead single particle electrode: Part I. graphitized carbon. *Electrochimica Acta*, 47(6):885 – 890, 2001.

[115] M. Holzapfel, A. Martinent, F. Alloin, B. Le Gorrec, R. Yazami, and C. Montella. First lithiation and charge discharge cycles of graphite materials, investigated by electrochemical impedance spectroscopy. *Journal of Electroanalytical Chemistry*, 546(0):41 – 50, 2003.

[116] Tiehua Piao, Su-Moon Park, Shil-Hoon Doh, and Seong-In Moon. Intercalation of lithium ions into graphite electrodes studied by ac impedance measurements. *Journal of The Electrochemical Society*, 146(8):2794–2798, 1999.

[117] Miran Gaberscek, Joze Moskon, Bostjan Erjavec, Robert Dominko, and Janez Jamnik. The importance of interphase contacts in li ion electrodes: The meaning of the high-frequency impedance arc. *Electrochemical and Solid-State Letters*, 11(10):A170–A174, 2008.

[118] E. Markevich, M.D. Levi, and D. Aurbach. Comparison between potentiostatic and galvanostatic intermittent titration techniques for determination of chemical diffusion coefficients in ion-insertion electrodes. *Journal of Electroanalytical Chemistry*, 580(2):231 – 237, 2005.

[119] A.V. Churikov, A.V. Ivanishchev, I.A. Ivanishcheva, V.O. Sycheva, N.R. Khasanova, and E.V. Antipov. Determination of lithium diffusion coefficient in $LiFePO_4$ electrode by galvanostatic and potentiostatic intermittent titration techniques. *Electrochimica Acta*, 55(8):2939 – 2950, 2010.

[120] Jan Philipp Schmidt, Philipp Berg, Michael Schönleber, André Weber, and Ellen Ivers-Tiffée. The distribution of relaxation times as basis for generalized time-domain models for li-ion batteries. *Journal of Power Sources*, 221(0):70 – 77, 2013.

[121] Wladislaw Waag, Stefan Käbitz, and Dirk Uwe Sauer. Experimental investigation of the lithium-ion battery impedance characteristic at various conditions and aging states and its influence on the application. *Applied Energy*, 102(0):885 – 897, 2013. Special Issue on Advances in sustainable biofuel production and use - {XIX} International Symposium on Alcohol Fuels - ISAF.

[122] D. Andre, M. Meiler, K. Steiner, Ch. Wimmer, T. Soczka-Guth, and D.U. Sauer. Characterization of high-power lithium-ion batteries by electrochemical impedance spectroscopy. i. experimental investigation. *Journal of Power Sources*, 196(12):5334 – 5341, 2011.

[123] D. Andre, M. Meiler, K. Steiner, H. Walz, T. Soczka-Guth, and D.U. Sauer. Characterization of high-power lithium-ion batteries by electrochemical impedance spectroscopy. II: Modelling. *Journal of Power Sources*, 196(12):5349 – 5356, 2011.

[124] G. Nagasubramanian. Two- and three-electrode impedance studies on 18650 li-ion cells. *Journal of Power Sources*, 87(1):226 – 229, 2000.

[125] Qunwei Wu, Wenquan Lu, and Jai Prakash. Characterization of a commercial size cylindrical li-ion cell with a reference electrode. *Journal of Power Sources*, 88(2):237 – 242, 2000.

[126] G. Nagasubramanian and D.H. Doughty. 18650 li-ion cells with reference electrode and in situ characterization of electrodes. *Journal of Power Sources*, 150(0):182 – 186, 2005.

[127] D.P. Abraham, J. Liu, C.H. Chen, Y.E. Hyung, M. Stoll, N. Elsen, S. MacLaren, R. Twesten, R. Haasch, E. Sammann, I. Petrov, K. Amine, and G. Henriksen. Diagnosis of power fade mechanisms in high-power lithium-ion cells. *Journal of Power Sources*, 119(0):511 – 516, 2003.

[128] D.P. Abraham, E.P. Roth, R. Kostecki, K. McCarthy, S. MacLaren, and D.H. Doughty. Diagnostic examination of thermally abused high-power lithium-ion cells. *Journal of Power Sources*, 161(1):648 – 657, 2006.

[129] D.P. Abraham, J.L. Knuth, D.W. Dees, I. Bloom, and J.P. Christophersen. Performance degradation of high - power lithium-ion cells - electrochemistry of harvested electrodes. *Journal of Power Sources*, 170(2):465 – 475, 2007.

[130] D. P. Abraham, D. W. Dees, J. Christophersen, C. Ho, and A. N. Jansen. Performance of high-power lithium-ion cells under pulse discharge and charge conditions. *International Journal of Energy Research*, 34(2):190–203, 2010.

[131] M. Kassem and C. Delacourt. Postmortem analysis of calendar-aged graphite/LiFePO$_4$ cells. *Journal of Power Sources*, 235(0):159 – 171, 2013.

[132] Isabel Jiménez Gordon, Sylvie Grugeon, Aurélie Débart, Gwennaelle Pascaly, and Stéphane Laruelle. Electrode contributions to the impedance of a high-energy density li-ion cell designed for ev applications. *Solid State Ionics*, 237(0):50 – 55, 2013.

[133] Tommy Georgios Zavalis, Matilda Klett, Maria H. Kjell, Mårten Behm, Rakel Wreland Lindström, and Göran Lindbergh. Aging in lithium-ion batteries: Model and experimental investigation of harvested LiFePO$_4$ and mesocarbon microbead graphite electrodes. *Electrochimica Acta*, 110(0):335 – 348, 2013.

[134] Moses Ender, André Weber, and Ivers-Tiffée Ellen. Analysis of three-electrode setups for ac-impedance measurements on lithium-ion cells by FEM simulations. *Journal of The Electrochemical Society*, 159(2):A128–A136, 2011.

[135] EL-Cell. http://el-cell.com/products/test-cells/ecc-ref.

[136] C.H. Chen, J. Liu, and K. Amine. Symmetric cell approach and impedance spectroscopy of high power lithium-ion batteries. *Journal of Power Sources*, 96(2):321 – 328, 2001.

[137] Jörg Illig, Moses Ender, Thorsten Chrobak, Jan Philipp Schmidt, Dino Klotz, and Ellen Ivers-Tiffée. Separation of charge transfer and contact resistance in LiFePO$_4$-cathodes by impedance modeling. *Journal of The Electrochemical Society*, 159(7):A952–A960, 2012.

[138] J. C. Burns, L. J. Krause, Dinh-Ba Le, L. D. Jensen, A. J. Smith, Deijun Xiong, and J. R. Dahn. Introducing symmetric li-ion cells as a tool to study cell degradation mechanisms. *Journal of The Electrochemical Society*, 158(12):A1417–A1422, 2011.

[139] Bernard A. Boukamp. Interpretation of an inductive loop in the impedance of an oxygen ion conducting electrolyte metal electrode system. *Solid State Ionics*, 143(1):47–55, 2001. Festschrift to honor Prof. Stan Whittingham.

[140] S. B. Adler. Reference electrode placement in thin solid electrolytes. *Journal of The Electrochemical Society*, 149(5):E166–E172, 2002.

[141] Amine Touzi. Untersuchung eines 3-Elektrodensetups mittels Experimentalzellen und FEM-Simulationen. Master's thesis, KIT, 2012.

[142] F. La Mantia, C.D. Wessells, H.D. Deshazer, and Yi Cui. Reliable reference electrodes for lithium-ion batteries. *Electrochemistry Communications*, 31(0):141 – 144, 2013.

[143] Bogdan Brad. Untersuchung des nichtlinearen Verhaltens einer Lithium-Ionen-Batterie. Master's thesis, KIT, 2013.

[144] Jörg Illig, Thorsten Chrobak, Moses Ender, Jan Philipp Schmidt, Dino Klotz, and Ellen Ivers-Tiffée. Studies on $LiFePO_4$ as cathode material in li-ion batteries. *ECS Transactions*, 28(30):3–17, 2010.

[145] Jan Philipp Schmidt, Thorsten Chrobak, Moses Ender, Jörg Illig, Dino Klotz, and Ellen Ivers-Tiffée. Studies on $LiFePO_4$ as cathode material using impedance spectroscopy. *Journal of Power Sources*, 196(12):5342 – 5348, 2011.

[146] Jörg Illig, Thorsten Chrobak, Dino Klotz, and Ellen Ivers-Tiffée. Evaluation of the rate determining processes for $LiFePO_4$ as cathode material in lithium-ion-batteries. *ECS Transactions*, 33(29):3–15, 2011.

[147] Moses Ender, Jochen Joos, Thomas Carraro, and Ellen Ivers-Tiffée. Quantitative characterization of $LiFePO_4$ cathodes reconstructed by FIB/SEM tomography. *Journal of The Electrochemical Society*, 159(7):A972–A980, 2012.

[148] Hai Yen Tran, Giorgia Greco, Corina Täubert, Margret Wohlfahrt-Mehrens, Wolfgang Haselrieder, and Arno Kwade. Influence of electrode preparation on the electrochemical performance of $LiNi_{0.8}Co_{0.15}Al_{0.05}O_2$ composite electrodes for lithium-ion batteries. *Journal of Power Sources*, 210(0):276 – 285, 2012.

[149] Doron Aurbach and Arie Zaban. Impedance spectroscopy of lithium electrodes: Part 1. general behavior in propylene carbonate solutions and the correlation to surface chemistry and cycling efficiency. *Journal of Electroanalytical Chemistry*, 348(1):155 – 179, 1993.

[150] Marc Doyle, Jeremy P. Meyers, and John Newman. Computer simulations of the impedance response of lithium rechargeable batteries. *Journal of The Electrochemical Society*, 147(1):99–110, 2000.

[151] J. Y. Song, Y. Y. Wang, and C. C. Wan. Conductivity study of porous plasticized polymer electrolytes based on poly(vinylidene fluoride) a comparison with polypropylene separators. *Journal of The Electrochemical Society*, 147(9):3219–3225, 2000.

[152] Arie Zaban, Ella Zinigrad, and Doron Aurbach. Impedance spectroscopy of li electrodes. 4. a general simple model of the li-solution interphase in polar aprotic systems. *The Journal of Physical Chemistry*, 100(8):3089–3101, 1996.

[153] Jiali Liu, Rongrong Jiang, Xiaoya Wang, Tao Huang, and Aishui Yu. The defect chemistry of $LiFePO_4$ prepared by hydrothermal method at different ph values. *Journal of Power Sources*, 194(1):536 – 540, 2009.

[154] Janko Jamnik and Miran Gaberscek. Li ion migration at the interfaces. *MRS Bulletin*, 34:942–948, 12 2009.

[155] A. van Bommel and Ranjith Divigalpitiya. Effect of calendering $LiFePO_4$ electrodes. *Journal of The Electrochemical Society*, 159(11):A1791–A1795, 2012.

[156] Doron Aurbach, Ella Zinigrad, Yaron Cohen, and Hanan Teller. A short review of failure mechanisms of lithium metal and lithiated graphite anodes in liquid electrolyte solutions. *Solid State Ionics*, 148(3):405 – 416, 2002.

[157] D. Aurbach, I. Weissman, A. Zaban, and O. Chusid. Correlation between surface chemistry, morphology, cycling efficiency and interfacial properties of li electrodes in solutions containing different li salts. *Electrochimica Acta*, 39(1):51 – 71, 1994.

[158] E. Peled, D. Golodnitsky, and G. Ardel. Advanced model for solid electrolyte interphase electrodes in liquid and polymer electrolytes. *Journal of The Electrochemical Society*, 144(8):L208–L210, 1997.

[159] Moses Ender, Jochen Joos, Thomas Carraro, and Ellen Ivers-Tiffée. Three-dimensional reconstruction of a composite cathode for lithium-ion cells. *Electrochemistry Communications*, 13(2):166 – 168, 2011.

[160] Barbara Stiaszny, Jörg C. Ziegler, Elke E. Krauss, Jan P. Schmidt, and Ellen Ivers-Tiffée. Electrochemical characterization and post-mortem analysis of aged $LiMn_2O_4$-NMC / graphite lithium ion batteries part I: Cycle aging. *Journal of Power Sources*, 251:439 – 450, 2013.

[161] Y.-H. Chen, C.-W. Wang, X. Zhang, and A.M. Sastry. Porous cathode optimization for lithium cells: Ionic and electronic conductivity, capacity, and selection of materials. *Journal of Power Sources*, 195(9):2851 – 2862, 2010.

[162] Honghe Zheng, Jing Li, Xiangyun Song, Gao Liu, and Vincent S. Battaglia. A comprehensive understanding of electrode thickness effects on the electrochemical performances of li-ion battery cathodes. *Electrochimica Acta*, 71(0):258 – 265, 2012.

[163] Honghe Zheng, Li Tan, Gao Liu, Xiangyun Song, and Vincent S. Battaglia. Calendering effects on the physical and electrochemical properties of $Li[Ni_{1/3}Mn_{1/3}Co_{1/3}]O_2$ cathode. *Journal of Power Sources*, 208(0):52 – 57, 2012.

[164] P.R. Shearing, L.E. Howard, P.S. Jorgensen, N.P. Brandon, and S.J. Harris. Characterization of the 3-dimensional microstructure of a graphite negative electrode from a li-ion battery. *Electrochemistry Communications*, 12(3):374 – 377, 2010.

[165] Yuki Yamada, Yasutoshi Iriyama, Takeshi Abe, and Zempachi Ogumi. Kinetics of lithium ion transfer at the interface between graphite and liquid electrolytes: Effects of solvent and surface film. *Langmuir*, 25(21):12766–12770, 2009. PMID: 19856995.

[166] Doron Aurbach, Mikhail D. Levi, Elena Levi, and Alexander Schechter. Failure and stabilization mechanisms of graphite electrodes. *The Journal of Physical Chemistry B*, 101(12):2195–2206, 1997.

[167] K. Dokko, Y. Fujita, M. Mohamedi, M. Umeda, I. Uchida, and J.R. Selman. Electrochemical impedance study of li-ion insertion into mesocarbon microbead single particle electrode: Part II. disordered carbon. *Electrochimica Acta*, 47(6):933 – 938, 2001.

[168] Takeshi Abe, Hideo Fukuda, Yasutoshi Iriyama, and Zempachi Ogumi. Solvated li-ion transfer at interface between graphite and electrolyte. *Journal of The Electrochemical Society*, 151(8):A1120–A1123, 2004.

[169] Takeshi Abe, Fumihiro Sagane, Masahiro Ohtsuka, Yasutoshi Iriyama, and Zempachi Ogumi. Lithium-ion transfer at the interface between lithium-ion conductive ceramic electrolyte and liquid electrolyte - a key to enhancing the rate capability of lithium-ion batteries. *Journal of The Electrochemical Society*, 152(11):A2151–A2154, 2005.

[170] Doron Aurbach, Boris Markovsky, Gregory Salitra, Elena Markevich, Yossi Talyossef, Maxim Koltypin, Linda Nazar, Brian Ellis, and Daniella Kovacheva. Review on electrode-electrolyte solution interactions, related to cathode materials for li-ion batteries. *Journal of Power Sources*, 165(2):491 – 499, 2007.

[171] Rotem Marom, Ortal Haik, Doron Aurbach, and Ion C. Halalay. Revisiting LiClO$_4$ as an electrolyte for rechargeable lithium-ion batteries. *Journal of The Electrochemical Society*, 157(8):A972–A983, 2010.

[172] Toshiyuki Momma, Mariko Matsunaga, Daikichi Mukoyama, and Tetsuya Osaka. Ac impedance analysis of lithium ion battery under temperature control. *Journal of Power Sources*, 216(0):304 – 307, 2012.

[173] Tetsuya Osaka, Toshiyuki Momma, Daikichi Mukoyama, and Hiroki Nara. Proposal of novel equivalent circuit for electrochemical impedance analysis of commercially available lithium ion battery. *Journal of Power Sources*, 205(0):483 – 486, 2012.

Werkstoffwissenschaft @ Elektrotechnik /
Universität Karlsruhe, Institut für Werkstoffe der Elektrotechnik

Die Bände sind im Verlagshaus Mainz (Aachen) erschienen.

Band 1   Helge Schichlein
Experimentelle Modellbildung für die Hochtemperatur-Brennstoffzelle SOFC. 2003
ISBN 3-86130-229-2

Band 2   Dirk Herbstritt
Entwicklung und Optimierung einer leistungsfähigen Kathoden-struktur für die Hochtemperatur-Brennstoffzelle SOFC. 2003
ISBN 3-86130-230-6

Band 3   Frédéric Zimmermann
Steuerbare Mikrowellendielektrika aus ferroelektrischen Dickschichten. 2003
ISBN 3-86130-231-4

Band 4   Barbara Hippauf
Kinetik von selbsttragenden, offenporösen Sauerstoffsensoren auf der Basis von $Sr(Ti,Fe)O_3$. 2005
ISBN 3-86130-232-2

Band 5   Daniel Fouquet
Einsatz von Kohlenwasserstoffen in der Hochtemperatur-Brennstoffzelle SOFC. 2005
ISBN 3-86130-233-0

Band 6   Volker Fischer
Nanoskalige Nioboxidschichten für den Einsatz in hochkapazitiven Niob-Elektrolytkondensatoren. 2005
ISBN 3-86130-234-9

Band 7   Thomas Schneider
Strontiumtitanferrit-Abgassensoren. Stabilitätsgrenzen / Betriebsfelder. 2005
ISBN 3-86130-235-7

Band 8   Markus J. Heneka
Alterung der Festelektrolyt-Brennstoffzelle unter thermischen und elektrischen Lastwechseln. 2006
ISBN 3-86130-236-5

Band 9   Thilo Hilpert
Elektrische Charakterisierung von Wärmedämmschichten mittels Impedanzspektroskopie. 2007
ISBN 3-86130-237-3

Band 10   Michael Becker
           Parameterstudie zur Langzeitbeständigkeit von
           Hochtemperaturbrennstoffzellen (SOFC). 2007
           ISBN 3-86130-239-X

Band 11   Jin Xu
           **Nonlinear Dielectric Thin Films for Tunable Microwave
           Applications.** 2007
           ISBN 3-86130-238-1

Band 12   Patrick König
           **Modellgestützte Analyse und Simulation von stationären
           Brennstoffzellensystemen.** 2007
           ISBN 3-86130-241-1

Band 13   Steffen Eccarius
           Approaches to Passive Operation of a Direct Methanol Fuel Cell. 2007
           ISBN 3-86130-242-X

Fortführung als

**Schriften des Instituts für Werkstoffe der Elektrotechnik,
Karlsruher Institut für Technologie (1868-1603)**

bei KIT Scientific Publishing

Die Bände sind unter www.ksp.kit.edu als PDF frei verfügbar oder als Druckausgabe bestellbar.

Band 14   Stefan F. Wagner
           **Untersuchungen zur Kinetik des Sauerstoffaustauschs
           an modifizierten Perowskitgrenzflächen.** 2009
           ISBN 978-3-86644-362-4

Band 15   Christoph Peters
           **Grain-Size Effects in Nanoscaled Electrolyte and Cathode
           Thin Films for Solid Oxide Fuel Cells (SOFC).** 2009
           ISBN 978-3-86644-336-5

Band 16   Bernd Rüger
           **Mikrostrukturmodellierung von Elektroden für die
           Festelektrolytbrennstoffzelle.** 2009
           ISBN 978-3-86644-409-6

Band 17   Henrik Timmermann
           **Untersuchungen zum Einsatz von Reformat aus flüssigen Kohlen-
           wasserstoffen in der Hochtemperaturbrennstoffzelle SOFC.** 2010
           ISBN 978-3-86644-478-2